# Analysis of Energy Efficiency: Perspectives and Policies towards Sustainable Development

# Analysis of Energy Efficiency: Perspectives and Policies towards Sustainable Development

Guest Editor

**Daniela Cristina MOMETE**

MDPI

Basel • Beijing • Wuhan • Barcelona • Belgrade • Novi Sad • Cluj • Manchester

*Guest Editor*
Daniela Cristina MOMETE
Faculty of Entrepreneurship
Business Engineering and
Management
National University of
Science and Technology
Politehnica Bucharest
Bucharest
Romania

*Editorial Office*
MDPI AG
Grosspeteranlage 5
4052 Basel, Switzerland

This is a reprint of the Special Issue, published open access by the journal *Energies* (ISSN 1996-1073), freely accessible at: https://www.mdpi.com/journal/energies/special_issues/6ZBVY6U564.

For citation purposes, cite each article independently as indicated on the article page online and as indicated below:

| Lastname, A.A.; Lastname, B.B. Article Title. *Journal Name* **Year**, *Volume Number*, Page Range. |
|---|

**ISBN 978-3-7258-6842-1 (Hbk)**
**ISBN 978-3-7258-6843-8 (PDF)**
**https://doi.org/10.3390/books978-3-7258-6843-8**

# Contents

# About the Editor

**Daniela Cristina MOMETE**

Daniela Cristina Momete has a remarkable academic background, with Bachelor of Science, Master of Science, and Ph.D. degrees in Chemical Engineering from the University Politehnica of Bucharest (now the National University of Science and Technology Politehnica Bucharest), earned in 1996, 1997, and 2004, respectively. She further diversified her expertise with a Ph.D. in International Business and Economics from the Bucharest Academy of Economic Studies. Her post-doctoral research at the Romanian Academy's National Institute for Economic Research concluded successfully in 2013, following several research fellowships across the U.K., France, and Italy.

Professor Momete's academic career at the University Politehnica of Bucharest spans over more than two decades. She served as a faculty member from 1997 to 2017 in the Department of Economic Engineering, progressing from teaching assistant to associate professor, and has been a full professor since 2017 in the same department within the Faculty of Entrepreneurship Business Engineering and Management. Her dedication to fostering new talent has been evident in her role as a Ph.D. coordinator in engineering management since 2019. She has published over 80 research papers and has contributed to 10 books. Since 2024, her expertise has extended globally through teaching and training programs in various European countries including Belgium, Germany, and the Baltic and Balkan states.

Her research interests are broad, encompassing energy economics, sustainable development, engineering education, entrepreneurship, and European economics. She has significantly contributed to the academic community by chairing numerous international conferences and serving as a reviewer for prominent conferences and journals, including MDPI. Professor Momete is an active member of leading professional organizations like the IEEE and the World Energy Council-R, and currently serves as an Associate Editor for *IEEE Access*.

# Preface

This Special Issue, "Analysis of Energy Efficiency: Perspectives and Policies towards Sustainable Development", serves as a platform to explore the subject of energy efficiency in depth, examining its scope across various sectors. Its primary aims are to synthesize current knowledge, identify emerging trends, and highlight innovative approaches in energy efficiency. Its purpose is to provide a comprehensive and rigorous analysis of both the technical and socio-economic dimensions of energy efficiency, linking them intimately with global sustainability goals.

Our planet faces unprecedented pressures from a growing population, industrialization, and increased consumption, all of which contribute to a relentless demand for energy. Historically, this demand has been met mainly through fossil fuels, leading to significant greenhouse gas emissions and geopolitical instabilities. The consequences, from diminishing natural resources to air, soil, and water pollution, emphasize the pressing need for a paradigm shift in how energy is produced, consumed, and managed.

I believe that this Special Issue, "Analysis of Energy Efficiency: Perspectives and Policies towards Sustainable Development", serves as a valuable resource for researchers, policymakers, industry leaders, and students alike. It offers a comprehensive overview of the current state of knowledge, identifies emerging trends, and highlights areas for further investigation. The diverse range of methodologies, case studies, and theoretical frameworks presented here will undoubtedly stimulate new research questions and foster interdisciplinary collaborations. It is my hope that this Special Issue will contribute meaningfully to the ongoing global dialogue on sustainable energy, inspire innovative solutions, and catalyze the necessary actions to build a more energy-efficient, equitable, and sustainable world. I extend my deepest gratitude to all the authors for their insightful contributions and to the reviewers for their meticulous efforts in ensuring the high quality and academic rigor of this publication.

**Daniela Cristina MOMETE**
*Guest Editor*

energies

MDPI

*Editorial*

# New Advances in Energy Efficiency: Perspectives and Policies Towards Sustainable Development

**Daniela Cristina Momete**

Department of Economic Engineering, Faculty of Entrepreneurship, Business Engineering and Management, National University of Science and Technology Politehnica Bucharest, Spl. Independentei, No. 313, S6, 060042 Bucharest, Romania; daniela.momete@upb.ro

## 1. Introduction

In the quest towards finding new sources of energy to ensure stable, sufficient, and affordable energy, we must consider "the most universally available source of energy" [1], i.e., energy efficiency must be considered. The measures taken to improve energy efficiency in modern times, which are characterized by a rising global energy demand, are meant to lower emissions, contributing to addressing climate change [2], with major impacts on human health, the environment, and global wellbeing [3], thereby enhancing overall sustainability [4]. A sustainable energy system based on efficiency measures implemented rationally and in a timely manner will lead to a more competitive economy and a safer and more prosperous life. New advances in technology and policies aimed at improving energy efficiency are essential for the transition to a sustainable future.

The European Union (EU) has set several targets for energy efficiency in the last few years. In 2021, EU legislation set very optimistic targets for energy, considering reductions of 36% and 39% in final and primary energy consumption, respectively, translating into a 9% increase in energy efficiency by 2030 [5]. Moreover, in 2022, the binding target was modified to a 13% increase in energy efficiency by 2030 [6] and then reset to an at least 11.7% increase for 2030 through the involvement of European legislators [7]. However, given the complexity of the energy system and the way in which the energy chain is interwoven with industry and the global economy, the pace of change and the implementation of measures must be tackled carefully as they can interfere with industrial competitiveness. In 2024, the European Economic and Social Committee released an opinion on the very ambitions climate targets and high investments needed for energy efficiency, potentially leading to higher costs for industries and threatening the competitiveness of EU industries [8].

Scientists are continuously looking for new methods of improving energy efficiency [9] to ensure the implementation of the sustainable development goals (SDGs) of Agenda 2030 [3], and a search on the Dimensions database [10] performed on 5 December 2024 revealed major interest in this area. The keywords included "energy efficiency" AND "sustainable development", mentioned over the last 10 years, and full data (comprising titles, keywords, abstracts, and text) were retrieved. There were 183,435 publications between 2014 and 2023, with the number of publications increasing from 9578 worldwide in 2014 to 39,751 in 2023, an increase of over 400%. Many of these papers, namely, 3809, were published in *Energies* (an MDPI journal), with 14 articles in 2014 and 892 articles in 2023. The academic impact of the papers published in *Energies*, expressed as citations, also revealed a major increase, rising from 3 citations in 2014 to 19,003 citations in 2023, with a projected 22,356 citations in 2024 (as some studies are still in progress), totaling 71,252 citations, indicating a major interest in energy efficiency and its importance for sustainable development.

 https://doi.org/10.3390/en18051171

The purpose of this manuscript is to provide an outline of the latest results and challenges in shaping the progress in energy efficiency, all stemming from studies submitted to this Special Issue, "Analysis of Energy Efficiency: Perspectives and Policies towards Sustainable Development".

## 2. Review of New Advances

Energy efficiency is vital for global energy systems. In this vein, the cheapest and most abundant form of energy is critical, with impacts on reducing the consumption of resources and lessening the strain of dependency on imported fossil fuels, offering solutions for better coping with increased energy prices and mitigating climate change.

Mornete [11] focused on the performance of EU-27 member states on the road towards an energy-efficient future, providing a comprehensive yet straightforward consumption effectiveness opportunity framework. The main findings revealed that only 19% of the EU-27 member states are active performers in regard to energy efficiency, most of which are located in Northen Europe, while the rest need to make large improvements and take measures specifically tailored to national contexts. The research indicated the fundamental importance of energy efficiency for reducing consumption, dependency on fossil fuels, and emissions, thus contributing to climate change mitigation. The main findings are valuable for researchers, legislators, and stakeholders in enhancing practices based on energy efficiency.

Abolhassani et al. [12] focused on energy intensity and studied the key performance indicators for energy and labor intensity by applying two novel statistical methods. They found a strong statistical correlation (0.67) between energy intensity and labor intensity in North American automotive plants. They also identified 11 factors and suggested that strategies aimed at enhancing energy efficiency can also improve labor productivity, and vice versa.

Ahn et al. [13] investigated the influence of initial pressure levels (starting from 1 bar to 3 bar) on compressed-air energy storage (CAES), an important commercial bulk-scale energy storage option. The main findings revealed that the greater the initial pressure, the lower the isothermal efficiency, leading to a decrease in air volume and mass, which, in turn, leads to a less-than-expected energy output. Their research indicated the importance of air dissolution, emphasizing that the changes in gas solubility depending on pressure are critical for understanding energy output during the expansion phase.

Murphy et al. [14] investigated the factors influencing the cost of saved electrical energy in customer-funded energy efficiency programs across most US states using an econometric model based on data reported by efficiency program administrators. Their research highlighted significant variations in cost and scale in various US states, indicating the importance of economies of scale, suggesting that the efficiency programs can be expanded without significantly increasing the costs in states where the efficiency savings were lower compared to retail sales. Their research presents valuable insights into the potential of energy efficiency as a valuable resource, offering improvements for policymakers, regulators, and system operators that forecast efficiency.

Benabdelkader et al. [15] proposed a control scheme for enhancing power quality in grid-connected single-phase photovoltaic (PV) systems. They designed an orthogonal signal generator for grid synchronization, and the results of the simulations and experiments showed better power quality and robustness against fluctuations. The proposed system used an adaptive neuro-fuzzy interference system to manage the voltage during faults at the point of common coupling. The effectiveness of this control strategy was assessed through simulations in MATLAB and experiments conducted on a small laboratory-scale photovoltaic system, which showed that the use of the PV system is beneficial for power quality.

Bigioi et al. [16] studied the financial performance of 43 energy companies based in Romania and analyzed the correlation between two variables: their governance and financial performance. The financial data were collected from the companies' financial statements, and the governance indicators were assessed using a scoring method. The main findings indicated a moderate intensity of the two variables, suggesting that the companies' effective governance systems aided them in achieving better performance.

Cuadros et al. [17] assessed the determinants of electrical energy generation efficiency in 24 Latin American and Caribbean countries for the 2000–2020 period by using Cragg's regression model. The main findings showed different results for efficient and inefficient countries. While tariff protectionism positively affected both efficient and inefficient countries, opening of the electricity sector positively impacted the efficient countries and market liberalization, and private power plant ownership improved efficiency in inefficient countries. These inefficient countries have a chance to improve their electricity generation efficiency by learning from the best practices of the efficient nations. The results of this study may be of interest to policymakers and regulators seeking valuable elements to improve efficiency in electricity generation, including private investment and policies regarding the liberalization of the electricity market.

Selim et al. [18] focused on the use of a sustainable energy source based on kinetic energy by exploring the use of piezoelectric sensors present in floor tiles to harvest energy from human footsteps. Piezoelectric sensors were installed beneath floor tiles, and a simple rectifying circuit with a filter was employed to harvest the electrical energy generated by the sensors to power two light-emitting diodes. The maximum power from the floor tile reached 249.6 milliwatts, at an estimated cost of implementation of about 10 USD/tile. The results suggest that a floor based on 30 strategically located floor tiles may generate 60 watt-hours of energy over 8 h.

## 3. Conclusions

The contributions to this Special Issue, "Analysis of Energy Efficiency: Perspectives and Policies towards Sustainable Development", demonstrate the progress in energy efficiency on several fronts: technological, political, economic, and social. The authors presented critical theoretical and practical aspects of energy efficiency and provided many valuable insights into the importance of energy efficiency for a sustainable future society.

The readers of this Special Issue may find interesting perspectives and technical solutions that help to unravel some of the challenges in energy efficiency. This Special Issue will be useful for policymakers, researchers, and students interested in engineering as well as economics. This collection of articles will allow readers to enrich their knowledge of energy efficiency, helping them better understand the complexity of this field as well as find inspiration to pursue further research.

**Funding:** This research received no external funding.

**Acknowledgments:** The author thanks the contributors to this Special Issue, "Analysis of Energy Efficiency: Perspectives and Policies towards Sustainable Development", for their valuable research. The author is grateful for the administrative and technical support provided by the *Energies* Editorial Office.

**Conflicts of Interest:** The authors declare no conflict of interest.

## References


1. European Commission. Clean Energy for All Europeans. Communication from the Commission to the European Parliament, The Council, The European Economic and Social Committee, The Committee of the Regions and the European Investment Bank, Brussels, COM(2016) 860 Final, p. 4. 2016. Available online: https://eur-lex.europa.eu/resource.html?uri=cellar:fa6ea15b-b7b0-11e6-9e3c-01aa75ed71a1.0001.02/DOC_1&format=PDF#:~:text=Energy%20efficiency%20is%20the%20most,to%20be%20produced%20or%20used (accessed on 1 December 2024).
2. United Nations. Paris Agreement. 2015. Available online: https://unfccc.int/files/essential_background/convention/application/pdf/english_paris_agreement.pdf (accessed on 11 November 2024).
3. United Nations (UN). Transforming Our World: The 2030 Agenda for Sustainable Development. 2015. Available online: https://documents-dds-ny.un.org/doc/UNDOC/GEN/N15/291/89/PDF/N1529189.pdf?OpenElement (accessed on 12 November 2024).
4. Momete, D.C. A Critical Analysis of the Primary Energy Consumption Trends from a Sustainable Perspective. *Univ. Politeh. Buchar. Sci. Bull. Ser. B-Chem. Mater. Sci.* **2007**, *69*, 73–80.
5. European Commission. *European Green Deal—Delivering on Our Targets*; Publications Office of the European Union: Luxembourg, 2021.
6. European Commission. REPowerEU Plan. Communication from The Commission to The European Parliament, The European Council, The Council, The European Economic and Social Committee and The Committee of the Regions, Brussels. COM(2022) 230 Final. 2022. Available online: https://eur-lex.europa.eu/legal-content/EN/TXT/HTML/?uri=CELEX:52022DC0230 (accessed on 1 December 2024).
7. European Parliament. Directive (EU) 2023/1791 of the European Parliament and of the Council of 13 September 2023 on Energy Efficiency and Amending Regulation (EU) 2023/955. 2023. Available online: https://eur-lex.europa.eu/legal-content/EN/TXT/PDF/?uri=CELEX:32023L1791 (accessed on 5 December 2024).
8. European Economic and Social Committee, Opinion on the Future of EU Industry in the Face of High Energy Prices and Transition Costs. 2024. Available online: https://www.eesc.europa.eu/en/our-work/opinions-information-reports/opinions/future-eu-industry-face-high-energy-prices-and-transition-costs (accessed on 7 December 2024).
9. Momete, D.C. Assessing the Complexity of Decarbonised Power Systems as a Means Towards a Sustainable Society: Trends and Hotspots. *Eur. J. Sustain. Dev.* **2024**, *13*, 411. [CrossRef]
10. Digital Science. Dimensions [Software]. 2018. Available online: https://app.dimensions.ai (accessed on 5 December 2024).
11. Momete, D.C. Salient Insights on the Performance of EU Member States on the Road towards an Energy-Efficient Future. *Energies* **2023**, *16*, 925. [CrossRef]
12. Abolhassani, A.; Boyd, G.; Jaridi, M.; Gopalakrishnan, B.; Harner, J. "Is Energy That Different from Labor?" Similarity in Determinants of Intensity for Auto Assembly Plants. *Energies* **2023**, *16*, 1776. [CrossRef]
13. Ahn, B.; Ro, P.I. Experimental Investigation of Impacts of Initial Pressure Levels on Compression Efficiency and Dissolution in Liquid Piston Gas Compression. *Energies* **2023**, *16*, 1921. [CrossRef]
14. Murphy, S.; Frick, N.M. Estimating the Drivers of the Cost of Saved Electricity in Utility Customer-Funded Energy Efficiency Programs. *Energies* **2023**, *16*, 2177. [CrossRef]
15. Benabdelkader, A.; Draou, A.; AlKassem, A.; Toumi, T.; Denai, M.; Abdelkhalek, O.; Ben Slimene, M. Enhanced Power Quality in Single-Phase Grid-Connected Photovoltaic Systems: An Experimental Study. *Energies* **2023**, *16*, 4240. [CrossRef]
16. Bigioi, A.D.; Bigioi, C.E. Governance and Performance in Romanian Energy Companies. *Energies* **2023**, *16*, 5041. [CrossRef]
17. Cuadros, S.; Rodríguez, Y.E.; Contreras, J. Determinants of the Efficiency of Electricity Generation in Latin America and Caribbean Countries Using a Cragg's Regression Model. *Energies* **2023**, *16*, 7875. [CrossRef]
18. Selim, K.K.; Smaili, I.H.; Yehia, H.M.; Ahmed, M.M.R.; Saleeb, D.A. Piezoelectric Sensors Pressed by Human Footsteps for Energy Harvesting. *Energies* **2024**, *17*, 2297. [CrossRef]

*energies*

*Article*

# Salient Insights on the Performance of EU Member States on the Road towards an Energy-Efficient Future

**Daniela Cristina Momete**

Department of Economic Engineering, Faculty of Entrepreneurship, Business Engineering and Management, University Politehnica of Bucharest, Spl. Independentei, No. 313, S6, 060042 Bucharest, Romania; daniela.momete@upb.ro

**Abstract:** Energy efficiency plays an important role in energy systems all over the world, representing the most available and cheapest source of energy. Progress in energy efficiency translates into reduced consumption, reduced dependency on external suppliers of fossil fuels, mitigation of climate change, and improvement of quality of life. Given its tremendous importance for energy systems and national economies, this paper aims to offer salient insights on the performance of the countries of the European Union (EU-27) towards an energy-efficient future, bearing in mind the quantitative milestones of sustainable development goal 7 (SDG7) of the Agenda for Sustainable Development of the United Nations. The current research offers a new perspective and delivers a comprehensive, robust, and succinct model which provides insights on the performance of EU-27 member states based on appropriate quantitative key performance indicators related to energy efficiency. The results show that about 19% of EU-27 member states are in a leading position, while about 41% of EU-27 member states have modest performance, and about 14% are still lagging. The findings of this paper may open discussions to find better ways to valorise energy efficiency, and may offer support for legislators and other stakeholders in energy systems. Moreover, by benchmarking different types of national practices, the countries which are still struggling to use energy-efficiency measures may find suitable ways to accelerate transition to a more energy-efficient future.

**Keywords:** energy efficiency; sustainable development; policy; prosperity; European Union; Agenda 2030

## 1. Introduction

The Agenda for Sustainable Development adopted by the United Nations (UN) in 2015 contains 17 sustainable development goals (SDGs) to be reached by 2030 [1]. Affordable and clean energy was formulated as a stand-alone goal (SDG7), but is also linked with other SDGs (SDG9, industry, innovation, and infrastructure; SDG11, sustainable cities and communities; SDG12, responsible consumption and production; SDG13, climate action, etc.).

SDG7 has four milestones which are synergically linked, which have not yet been accomplished, although there are only eight years to the target year:

- 7.1: "universal access to affordable, reliable and modern energy services" [1]. Unfortunately, 733 million people still have no access to electrical energy and 2.4 billion have no access to clean cooking areas [2], and there are few chances to achieve this goal by 2030.
- 7.2: "increase substantially the share of renewable energy in the global energy mix" [1]. The pace of renewable energy should increase in total final energy consumption if this goal is to be correlated with energy emissions belonging to SDG13, which, presently, seems out of reach.
- 7.3: "double the global rate of improvement in energy efficiency" [1] in comparison with the period 1990–2010. The primary energy intensity should increase in the

remaining years till 2030, with 3.2% per year, and this is still far from the actual 1.9% achieved [2].

- 7.A: "enhance international cooperation to facilitate access to clean energy research and technology", and 7.B: "expand infrastructure and upgrade technology for supplying modern and sustainable energy services for all in developing countries" [1,2]. The achievement of these targets, even though they are qualitatively formulated, requires more investment from public and private sectors, which, instead of increasing, dropped within recent years [2].

Among the milestones above listed, only one is quantitatively formulated (goal 7.3), referring to energy efficiency and its tremendous potential, and represents the key to improve wellbeing around the world. Energy efficiency should be improved in all aspects, from production to consumption, to reduce consumption, reduce the national dependency on external suppliers of fossil fuels, mitigate climate change. and improve quality of life. In this respect, the European Union (EU) set very ambitious targets for 2030 of 36% and 39% reduction in final and primary energy consumption, respectively. This translates into final and primary energy consumption of 787 Mtoe and 1023 Mtoe, respectively, by 2030 [3]. By 2050, the EU has an ambition to be "the first climate neutral continent" [3] and to reduce emissions by 50% compared with those registered in 1990. Major energy access programs were developed to succeed in milestone 7.1, but less effort was devoted to the efficient and productive use of energy [4].

Energy efficiency was defined by EU as "the ratio of output of performance, service, goods or energy, to input of energy" [5]. Energy efficiency represents the cheapest energy source, and most available one, for all countries around the globe, being considered "the most cost-effective means to supply energy in a sustainable manner" [6]. Many studies discussed the benefits of energy efficiency and its long-term advantages [7,8], but with no major results.

Energy efficiency was the subject of a specific EU directive adopted in 2012, which acknowledged that improvement in this area should be at least 20% by 2020 [5]. In 2018, this target was revised to at least 32.5% improvement, but the target year was changed to 2030. Moreover, the European Green Deal, with its ambitious goal to reduce greenhouse gases emissions by 55% by 2050 [3], led to further revision of the energy efficiency directive in 2021 [9]. Although the energy efficiency target of 20% by 2020 was achieved at the EU level in 2020 due to circumstances of COVID-19, there is a major gap between the national contributions communicated by the member states and the ambition of 32.5% by 2030 [9,10].

While energy savings are considered to be the greatest opportunity, there are major challenges, especially in energy losses in energy transformation, heating and cooling, the transport sector, buildings, and, lately, the information and communication sector (ITC) [11,12].

At the global level, a plethora of policies have been thought of for energy efficiency, and more than 2800 national policies are already in force in different countries, with 1383 in European countries issued within the period 1951–2022 [13]. The story of energy efficiency policies in Europe started with the electricity tax in Norway in 1951, while the most recent is the France 2030 Investment Plan issued by France in 2022 [13]. For Europe, the EU created an energy dashboard [14], but energy efficiency is judged only in terms of consumption of energy, without considering the innovation aspect (the possibility of creating opportunities through R&D and patents related to energy efficiency). Moreover, the EU created a tool to monitor a portion of energy-efficiency measurements—energy savings [15]—but developed no common methodology for the evaluation and measurement of these savings. A devoted family of standards for energy management was created by the ISO organization through its ISO 50001 [16], but is not yet widely used.

Energy efficiency requires technological solutions, but also better management and motivation of consumers [17]. Powerful ways to lower energy usage, such as higher efficiency engines, increased heat recovery, better process design, and new breakthrough technology, are a necessity, but not a sufficient solution on their own. This sufficiency can be acquired by also taking into consideration the behaviour of businesses, individual consumers, and the state. For corporations, the journey to zero-economy and energy transition brings additional pressure concerning their competitiveness. Despite the high potential of energy-efficiency measures, an energy paradox has been shown by many researchers [18,19], pointing that there is a large gap between theoretical cost-effective energy-efficient measures and actually implemented measures. The barriers to energy efficiency, which are undermining the energy-efficiency goals and measures, vary from economic (cost of production disruption, lack of time, and lack of capital), to institutional (lack of regulations and limitations of infrastructure), to behavioural (lack of culture of savings, inertia, and mistrust) [20,21]. The barriers to energy efficiency must be broken down by a transformative change, and this has, as its core, the revision of how performance, or progress, is measured.

Despite the pledges on sustainable transformation of economies, the focus is still on gross domestic product, and ever-increasing production and consumption fails to consider the improvements in efficient resources use. Furthermore, the extreme events correlated with climate change, pandemics, and military conflicts will determine a surge in the use of fossil fuels, especially of coal, gas (in gaseous and liquid form), and nuclear fuel [22–27]. Moreover, in energy efficiency and climate-friendly investments, better decision-support tools are a necessity, and they must be based on transparent metrics and ranking [28].

The Internet of Thing concept (IoT) refers, in brief, to the things which are connected to automatically communicate, the aim being to provide services for human benefit [29]. The IoT concept has recently been connected with energy-related problems to monitor energy consumption and prevent energy losses [30]. Digitalization offers new possibilities to enhance energy efficiency, and the IoT provides the opportunity to decrease energy consumption. The devices used to monitor energy consumption (examples include sensors for temperature, electricity, solar incidence, and humidity) helped to achieve an important reduction in energy consumption [31], and therefore must be considered when an analysis regarding energy efficiency is performed.

The aim of this research is to offer important insights on the assessment of the performance of EU-27 member states on their road towards an energy-efficient future by the creation and application of a decision-support tool based on a multidimensional and comprehensive framework. To this end, the most appropriate and available key performance indicators (KPIs) were selected and applied through a step-by-step methodology to create a metric and assess a ranking of performance within EU-27 member states. The assessment of energy efficiency performance calls for the identification and subsequent implementation of KPIs that help evaluate progress towards an energy-efficient future. Despite the relevance of the topic of energy efficiency, knowledge about the benefits of fully quantitative terms is scattered [32], and none is included the IoT, to the best knowledge of the author. Furthermore, several initiatives developed very overcrowded and complex models, or had a focus set on a qualitative scoring [33–35], which is difficult to apply and interpret, and several authors limited their analysis to specific industries [36].

This research aims to bridge the identified gaps through the proposal of an original, easy to understand and apply, yet comprehensible and multidimensional framework which offers a quintessence of energy efficiency. This framework is to be used for the assessment of the performance of the EU member states on the road towards an energy-efficient future. The contributions of this research are:

1. Identification of suitable and available quantitative KPIs relevant to energy efficiency, the focus being the quantitative representation of the milestones of SDG7.
2. Building of a comprehensive model which implies a robust, yet succinct, framework based on quantification (normalization, weighting depending on the conferred importance, and design of clusters on a balanced score).
3. Applying the model to the EU-27 to identify the status of each member state on the road towards an energy-efficient future (ranked from best to worst).

The rest of this paper is organized as follows: Section 2 describes the methodological steps of research, Section 3 presents the results, Section 4 is devoted to the discussion of results and future research directions, and Section 5 presents the concluding remarks.

## 2. Methodological Steps

One of the most powerful key measures aimed to transform the energy system into a more sustainable one is energy efficiency. This research aims to identify the proper ways of transforming the energy systems within the EU-27 into more efficient ones, by considering the actual quantitative outcomes of each member state.

The methodology has as its core the quantitative representation of the milestones of SDG7, and is based on an original design developed by the author. It aims to monitor consumption, with a focus on primary consumption and the use of the IoT by enterprises, to target effectiveness, with regards to energy intensity and the share of renewable energy sources (RES) used in final consumption, and to consider opportunities through access to technology by investment in research and development (R&D) and patents related to energy. In this way, the identification of the main gaps in progress towards an energy-efficient future moves beyond the required energy resources to effectiveness and opportunities, to reflect the needed effort.

This research develops an original framework developed on three factors, the Consumption-Effectiveness-Opportunities (CEO) framework (see Figure 1). The framework considers three factors, each with three KPIs, specifically designed to capture the quintessence of energy efficiency (matrix 3 × 2). Each factor carries specific key performance indicators (KPI), equally balanced to reflect the importance of each factor (see Table 1). The data were collected from internationally reliable centralized sources such as Eurostat and SETIS.

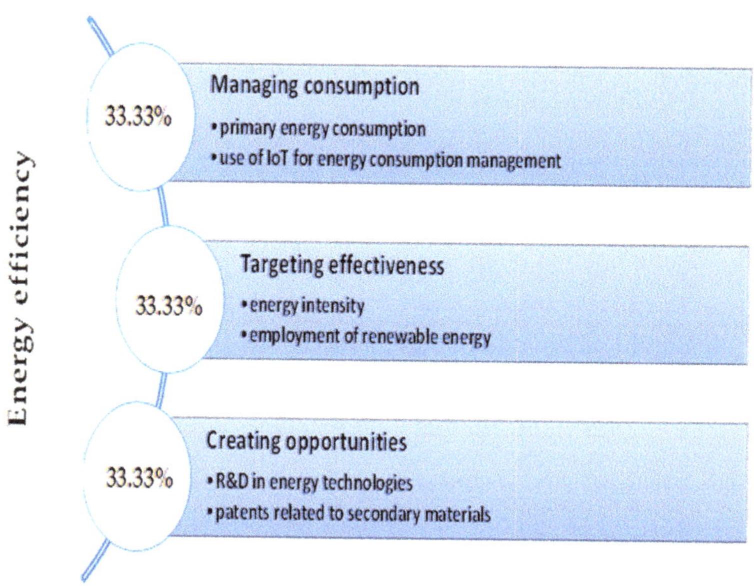


**Figure 1.** Consumption-Effectiveness-Opportunities (CEO) framework.

**Table 1.** Factors and KPIs for CEO framework.

| Factor | Code | KPI (Definition) | Unit | Reason of Selection | Effect on the Energy Efficiency |
|---|---|---|---|---|---|
| **Managing consumption** | I1 | Primary energy consumption (gross inland consumption of energy, excluding non-energy use) | toe/capita | An increase in energy consumption decreases the energy efficiency | negative |
| | I2 | Manufacturing enterprises which use IoT for energy consumption management (smart sensors for lighting, temperature, water etc, used by small, medium-sized and large enterprises) | percentage of all manufacturing enterprises | The presence of IoT in energy consumption management increases the energy efficiency | positive |
| **weight 33.33%** | | | | | |
| **Targeting effectiveness** | I3 | Energy intensity (the ratio of total energy supply to the annual gross domestic product) | Kgoe/1000 EUR-PPS | A decrease in energy intensity is generally linked with an increase in energy efficiency | negative |
| | I4 | Renewable energy sources in final energy consumption | percentage | An increase in RES usage in energy mix is generally linked with an enhancement of energy efficiency (especially for electricity) | positive |
| **weight 33.33%** | | | | | |
| **Creating opportunities** | I5 | R&D in energy technologies (investments in research and development both private and public) | Mil EUR/100,000 inhabitants | An increase in the investments in R&D in energy technologies offers the potential of an enhancement in energy efficiency | positive |
| | I6 | Patents related to secondary raw materials | Number/1,000,000 inhabitants | An increase in the number of patents for energy savings offers the potential of an enhancement in energy efficiency | positive |
| **weight 33.33%** | | | | | |

The three factors were considered equally important, as the best results will be obtained by synergically considering managing consumption, targeting effectiveness, and creating opportunities. The three factors, regardless of the number of KPIs considered, must consider the reconciliation of the needs of society in terms of energy consumption with sustainable economic growth. In this way, the current model moves away from the dominant global economic paradigm and does not subordinate energy efficiency to economic growth at all costs. Moreover, the original operational framework has in view that simplicity is a virtue in science [37–39], and considers comprehensive KPIs to accurately render the energy-efficiency quintessence, based on safe and rational consumption.

The CEO model consists of identification of a CEO composite index (ICEO) to be further used to identify the best and worst performances on the road towards an energy-efficient future. The ICEO brings together a mix of determinants which are mathematically linked to result in one single number which reflects the quintessence of energy efficiency.

The ICEO allows the identification of differences across the EU-27 space, and facilitates the identification of further interventions.

The CEO model implies five phases, and the logical scheme is presented in Figure 2:

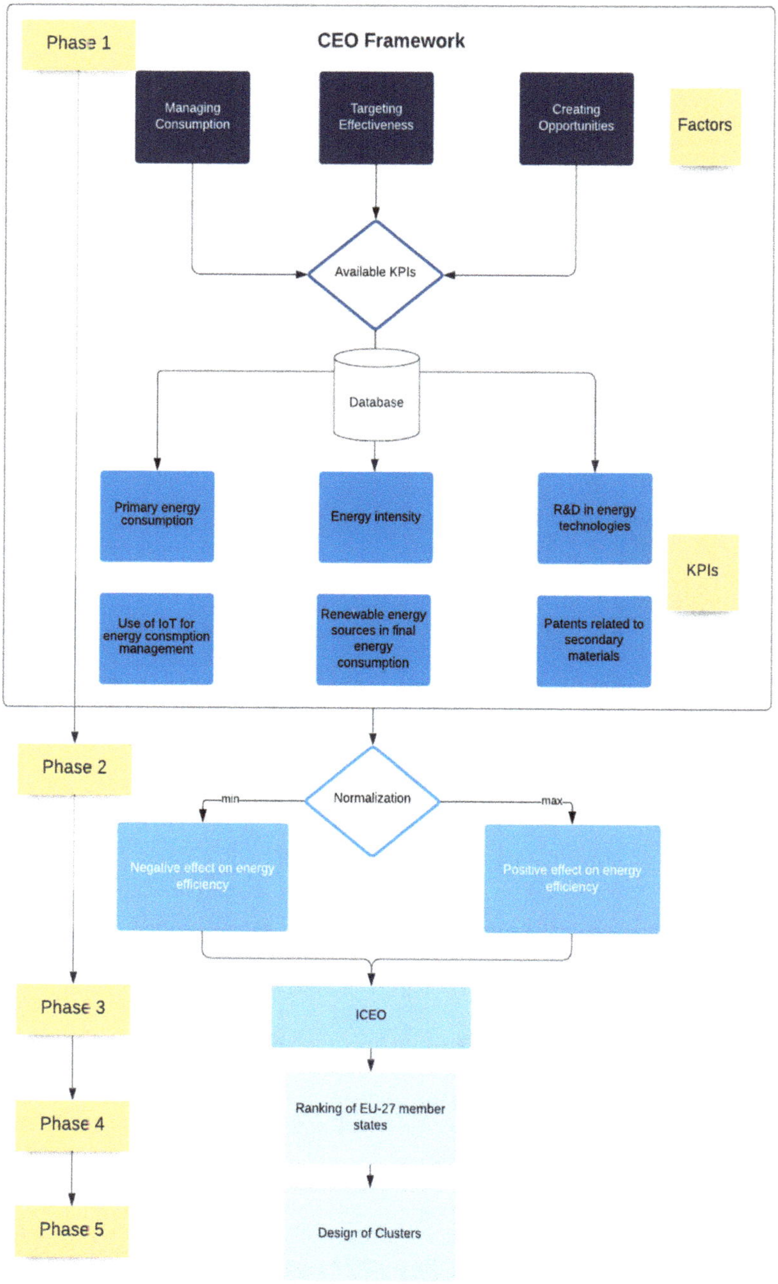


**Figure 2.** Logical scheme of the phases of research. Source: author's own representation using Lucidchart.

Phase 1. Design of the framework of the model. Selection of suitable and available indicators for each factor (matrix 3 × 2). Last available year for all countries and factors was employed, and the research does not include the effects of COVID-19.

Phase 2. Normalization of the KPIs, given the fact that they use different units. The population of each country was considered, to make possible comparisons among large and small countries [40]. The data were normalized by their transposition into dimensionless units to make them comparable [41] according to their effect on energy efficiency:

$$Positive\ effect:\ Nij = \frac{aij - ajmin}{ajmax - ajmin},\ i \in (1,\ 6),\ j \in (1,\ 27) \tag{1}$$

$$Negative\ effect:\ Nij = \frac{ajmax - aij}{ajmax - ajmin},\ i \in (1,\ 6),\ j \in (1,\ 27) \tag{2}$$

where $i$ = KPIs (6), $j$ = number of countries (27), $Nij$ = normalized value for factor for country $j$, and KPI $i$ ($0 \leq N \leq 1$).

Phase 3. Calculation of the CEO composite index (ICEO) which gives equal weigh to factors:

$$ICEO = \left[\frac{I1 + I2}{2}\right] \times 0.333 + \left[\frac{I3 + I4}{2}\right] \times 0.333 + \left[\frac{I5 + I6}{2}\right] \times 0.333 \tag{3}$$

Phase 4. Application of the model to EU-27 member states and organization of states from best to worst in terms of energy efficiency.

Phase 5. Design of four clusters, depending on the distance to the energy efficiency target, in increments of 0.1.

Cluster I: ICEO > 0.5

Cluster II: 0.4 < ICEO < 0.5

Cluster III: 0.3 < ICEO < 0.4

Cluster IV: ICEO < 0.3.

## 3. Results

The evolution of the EU-27 in terms of key energy efficiency indicators was evaluated within the interval of 2000–2020, and the results are shown in Table 2. Primary energy consumption decreased over the 2000–2020 interval by more than 15%, while final energy consumption decreased by 11%. In terms of energy intensity, the EU-27 acquired a total decrease of 52% within the 2000–2020 interval. A major contribution of the decrease in consumption might be attributed to the effects of COVID-19, as the values of energy consumption decreased sharply, by about 8%, from 2019 to 2020. Therefore, the data for the CEO model only included data up to 2019, to exclude the effects of COVID-19, which put forward an important disturbance of markets.

**Table 2.** Evolution of energy-efficiency indicators within the EU-27.

| EU-27 | Primary Energy Consumption, Toe/Capita [42] | Final Energy Consumption, Toe/Capita [43] | Energy Intensity, KGOE/1000 EUR in PPS [44] | Primary Energy, Yearly Variance, % | Final Energy, Yearly Variance, % | Energy Intensity, Yearly Variance, % |
|---|---|---|---|---|---|---|
| 2000 | 3.26 | 2.28 | 240.86 | 2.45 | 2.19 | −1.72 |
| 2001 | 3.34 | 2.33 | 236.72 | 0.00 | −0.86 | −5.03 |
| 2002 | 3.34 | 2.31 | 224.81 | 2.10 | 2.60 | −1.53 |
| 2003 | 3.41 | 2.37 | 221.38 | 0.88 | 0.84 | −4.85 |
| 2004 | 3.44 | 2.39 | 210.63 | 0.00 | 0.00 | −6.04 |
| 2005 | 3.44 | 2.39 | 197.90 | 0.58 | 0.42 | −5.41 |

**Table 2.** *Cont.*

| EU-27 | Primary Energy Consumption, Toe/Capita [42] | Final Energy Consumption, Toe/Capita [43] | Energy Intensity, KGOE/1000 EUR in PPS [44] | Primary Energy, Yearly Variance, % | Final Energy, Yearly Variance, % | Energy Intensity, Yearly Variance, % |
|---|---|---|---|---|---|---|
| 2006 | 3.46 | 2.40 | 187.20 | −1.73 | −2.08 | −6.84 |
| 2007 | 3.40 | 2.35 | 174.41 | −0.29 | 0.43 | −5.31 |
| 2008 | 3.39 | 2.36 | 165.14 | −5.90 | −5.51 | −0.97 |
| 2009 | 3.19 | 2.23 | 163.54 | 3.45 | 4.04 | 0.59 |
| 2010 | 3.30 | 2.32 | 164.51 | −2.73 | −3.45 | −5.34 |
| 2011 | 3.21 | 2.24 | 155.72 | −1.25 | −0.45 | −4.15 |
| 2012 | 3.17 | 2.23 | 149.26 | −1.26 | −0.45 | −3.36 |
| 2013 | 3.13 | 2.22 | 144.24 | −4.15 | −4.50 | −4.69 |
| 2014 | 3.00 | 2.12 | 137.48 | 1.67 | 1.89 | −3.59 |
| 2015 | 3.05 | 2.16 | 132.55 | 0.33 | 1.85 | 0.37 |
| 2016 | 3.06 | 2.20 | 133.04 | 1.31 | 0.91 | −1.47 |
| 2017 | 3.10 | 2.22 | 131.09 | −0.65 | 0.00 | −4.14 |
| 2018 | 3.08 | 2.22 | 125.66 | −1.62 | −0.45 | −5.60 |
| 2019 | 3.03 | 2.21 | 118.62 | −8.91 | −8.14 | −3.54 |
| 2020 | 2.76 | 2.03 | 114.42 | | | |
| Growth rate (%) | −15.30 | −11.00 | −52.50 | | | |
| Standard deviation | 0.18 | 0.10 | 39.16 | | | |
| Mean (2000–2020) | 3.22 | 2.27 | 168.06 | | | |
| Variation coefficient | 0.06 | 0.04 | 0.23 | | | |

The convergence analysis was also investigated for the EU-27, within the interval of 2000–2020. Statistical tools were employed to analyse the data sets: mean, standard deviation, and the variation coefficient, which indicates the size of standard deviation relative to the mean. The values of the variation coefficient indicate that convergence towards the mean was higher.

The phases presented in Figure 2 were applied, and the results are shown in Tables 3–5. Table 3 contains data about the KPIs, with values retrieved from [42,44–49]. The statistical tools employed showed that the average lies far from the mean in the case of I3 and I5, whereas the dispersion is higher around the mean for I4 and I6.

**Table 3.** KPIs of the factors of the CEO model.

| Country/Factor | I1 | I2 | I3 | I4 | I5 | I6 |
|---|---|---|---|---|---|---|
| Belgium | 4.21 | 10 | 151.55 | 6.96 | 9.93 | 0.91 |
| Bulgaria | 2.61 | 2 | 162.76 | 0.04 | 21.55 | 0.00 |
| Czechia | 3.72 | 9 | 138.05 | 2.10 | 16.24 | 0.90 |
| Denmark | 2.89 | 10 | 77.59 | 23.48 | 37.02 | 1.22 |
| Germany | 3.43 | 11 | 98.38 | 17.26 | 17.27 | 1.03 |
| Estonia | 3.55 | 6 | 145.77 | 0.08 | 31.73 | 0.00 |
| Ireland | 2.98 | 7 | 51.55 | 3.53 | 11.98 | 1.42 |
| Greece | 2.08 | 10 | 117.75 | 0.01 | 19.63 | 0.00 |
| Spain | 2.56 | 8 | 100.24 | 0.92 | 17.85 | 0.36 |

**Table 3.** *Cont.*

| Country/Factor | I1 | I2 | I3 | I4 | I5 | I6 |
|---|---|---|---|---|---|---|
| France | 3.49 | 5 | 112.72 | 6.59 | 17.17 | 0.58 |
| Croatia | 2.02 | 9 | 104.34 | 0.07 | 28.47 | 0.00 |
| Italy | 2.44 | 8 | 87.67 | 1.93 | 18.18 | 0.53 |
| Cyprus | 2.88 | 9 | 114.30 | 0.69 | 13.78 | 0.00 |
| Latvia | 2.38 | 7 | 119.01 | 0.31 | 40.93 | 0.00 |
| Lithuania | 2.25 | 7 | 109.15 | 0.36 | 25.48 | 0.36 |
| Luxembourg | 7.26 | 6 | 91.81 | 17.10 | 7.05 | 1.61 |
| Hungary | 2.51 | 5 | 119.87 | 0.28 | 12.63 | 0.10 |
| Malta | 1.73 | 12 | 194.60 | 0.81 | 8.23 | 0.00 |
| Netherlands | 3.66 | 12 | 125.66 | 6.88 | 8.89 | 1.22 |
| Austria | 3.63 | 19 | 99.13 | 16.23 | 33.76 | 1.11 |
| Poland | 2.64 | 5 | 121.83 | 0.40 | 15.38 | 0.40 |
| Portugal | 2.14 | 7 | 98.18 | 0.97 | 30.62 | 0.19 |
| Romania | 1.66 | 3 | 79.15 | 0.13 | 24.29 | 0.28 |
| Slovenia | 3.12 | 19 | 119.67 | 1.06 | 21.97 | 0.48 |
| Slovakia | 2.93 | 9 | 143.53 | 0.50 | 16.89 | 0.67 |
| Finland | 5.80 | 10 | 182.81 | 11.62 | 42.72 | 2.96 |
| Sweden | 4.45 | 12 | 135.64 | 10.33 | 55.79 | 1.01 |
| | | | Statistical tools | | | |
| Observations | 27 | 27 | 27 | 27 | 27 | 27 |
| Observations with missing data | 0 | 0 | 0 | 0 | 0 | 0 |
| Minimum | 1.66 | 19.00 | 51.00 | 0.01 | 7.05 | 0.00 |
| Maximum | 7.26 | 19.00 | 194.60 | 23.48 | 55.79 | 2.93 |
| Mean | 3.15 | 8.78 | 118.62 | 4.84 | 22.42 | 0.64 |
| Standard deviation | 1.20 | 3.86 | 31.22 | 6.61 | 11.69 | 0.67 |
| Coefficient of variation | 0.38 | 0.44 | 0.26 | 1.37 | 0.52 | 1.05 |

Sources: data processed from [42,44–49].

**Table 4.** Normalized values of the KPI of each factor of CEO model.

| Country/Utilities | N1 | N2 | N3 | N4 | N5 | N6 |
|---|---|---|---|---|---|---|
| Belgium | 0.545 | 0.471 | 0.301 | 0.296 | 0.059 | 0.307 |
| Bulgaria | 0.830 | 0.000 | 0.223 | 0.001 | 0.298 | 0.000 |
| Czechia | 0.632 | 0.412 | 0.395 | 0.089 | 0.189 | 0.304 |
| Denmark | 0.780 | 0.471 | 0.818 | 1.000 | 0.615 | 0.412 |
| Germany | 0.684 | 0.529 | 0.673 | 0.735 | 0.210 | 0.348 |
| Estonia | 0.663 | 0.235 | 0.341 | 0.003 | 0.506 | 0.000 |
| Ireland | 0.764 | 0.294 | 1.000 | 0.150 | 0.101 | 0.480 |
| Greece | 0.925 | 0.471 | 0.537 | 0.000 | 0.258 | 0.000 |
| Spain | 0.839 | 0.353 | 0.660 | 0.039 | 0.222 | 0.122 |
| France | 0.673 | 0.176 | 0.572 | 0.280 | 0.208 | 0.196 |
| Croatia | 0.936 | 0.412 | 0.631 | 0.003 | 0.439 | 0.000 |
| Italy | 0.861 | 0.353 | 0.748 | 0.082 | 0.228 | 0.179 |
| Cyprus | 0.782 | 0.412 | 0.561 | 0.029 | 0.138 | 0.000 |
| Latvia | 0.871 | 0.294 | 0.528 | 0.013 | 0.695 | 0.000 |
| Lithuania | 0.895 | 0.294 | 0.597 | 0.015 | 0.378 | 0.122 |
| Luxembourg | 0.000 | 0.235 | 0.719 | 0.728 | 0.000 | 0.544 |
| Hungary | 0.848 | 0.176 | 0.522 | 0.012 | 0.115 | 0.034 |

**Table 4.** *Cont.*

| Country/Utilities | N1 | N2 | N3 | N4 | N5 | N6 |
|---|---|---|---|---|---|---|
| Malta | 0.988 | 0.588 | 0.000 | 0.034 | 0.024 | 0.000 |
| Netherlands | 0.643 | 0.588 | 0.482 | 0.293 | 0.038 | 0.412 |
| Austria | 0.648 | 1.000 | 0.667 | 0.691 | 0.548 | 0.375 |
| Poland | 0.825 | 0.176 | 0.509 | 0.017 | 0.171 | 0.135 |
| Portugal | 0.914 | 0.294 | 0.674 | 0.041 | 0.484 | 0.064 |
| Romania | 1.000 | 0.059 | 0.807 | 0.005 | 0.354 | 0.095 |
| Slovenia | 0.739 | 1.000 | 0.524 | 0.045 | 0.306 | 0.162 |
| Slovakia | 0.773 | 0.412 | 0.357 | 0.021 | 0.202 | 0.226 |
| Finland | 0.261 | 0.471 | 0.082 | 0.495 | 0.732 | 1.000 |
| Sweden | 0.502 | 0.588 | 0.412 | 0.440 | 1.000 | 0.341 |

**Table 5.** ICEO, ranking and clusters of countries.

<table>
<tr><th>Country</th><th>ICEO</th><th>Gap (Difference between the Leader and Other Member States) %</th><th>Cluster</th></tr>
<tr><td>Denmark</td><td>0.683</td><td>0</td><td rowspan="5">Cluster I<br>Countries with significant performance (Active performers)<br>ICEO > 0.5<br>Cluster of 5 countries = 19% of EU-27</td></tr>
<tr><td>Austria</td><td>0.655</td><td>4</td></tr>
<tr><td>Sweden</td><td>0.547</td><td>25</td></tr>
<tr><td>Germany</td><td>0.530</td><td>29</td></tr>
<tr><td>Finland</td><td>0.507</td><td>35</td></tr>
<tr><td>Ireland</td><td>0.465</td><td>47</td><td rowspan="7">Cluster II<br>Countries with moderate performance (Moderate performers)<br>0.4 < ICEO < 0.5<br>Cluster of 7 countries = 26% of EU-27</td></tr>
<tr><td>Slovenia</td><td>0.463</td><td>48</td></tr>
<tr><td>Portugal</td><td>0.412</td><td>66</td></tr>
<tr><td>Italy</td><td>0.409</td><td>67</td></tr>
<tr><td>Netherlands</td><td>0.408</td><td>67</td></tr>
<tr><td>Croatia</td><td>0.404</td><td>69</td></tr>
<tr><td>Latvia</td><td>0.400</td><td>71</td></tr>
<tr><td>Romania</td><td>0.387</td><td>76</td><td rowspan="11">Cluster III<br>Countries with modest performance (Modest performers) 0.3 < ICEO < 0.4<br>Cluster of 11 countries = 41% of EU-27</td></tr>
<tr><td>Lithuania</td><td>0.384</td><td>78</td></tr>
<tr><td>Spain</td><td>0.373</td><td>83</td></tr>
<tr><td>Luxembourg</td><td>0.371</td><td>84</td></tr>
<tr><td>Greece</td><td>0.365</td><td>87</td></tr>
<tr><td>France</td><td>0.351</td><td>95</td></tr>
<tr><td>Czechia</td><td>0.337</td><td>103</td></tr>
<tr><td>Slovakia</td><td>0.332</td><td>106</td></tr>
<tr><td>Belgium</td><td>0.330</td><td>107</td></tr>
<tr><td>Cyprus</td><td>0.320</td><td>113</td></tr>
<tr><td>Poland</td><td>0.306</td><td>123</td></tr>
<tr><td>Estonia</td><td>0.291</td><td>135</td><td rowspan="4">Cluster IV<br>Countries with very modest performance (Slow performers)<br>ICEO < 0.3<br>Cluster of 4 countries = 14% of EU-27</td></tr>
<tr><td>Hungary</td><td>0.285</td><td>140</td></tr>
<tr><td>Malta</td><td>0.272</td><td>151</td></tr>
<tr><td>Bulgaria</td><td>0.225</td><td>204</td></tr>
</table>

For I1, the best performer was Romania, with 1.66 toe/capita of primary energy consumption, while the worst performer was Luxemburg, with 7.26 toe/capita of primary energy consumption. Luxemburg consumed more than four times the energy consumed by Romania (toe = tons of oil equivalent = 41.868 MJ).

For I2, the best performer was Slovenia, with 19% of its small, medium sized, and large manufacturing enterprises using the IoT for monitoring energy consumption, while the worst performer was Bulgaria, with only 2%. Slovenia was more than nine times more involved in the IoT in manufacturing enterprises than Bulgaria.

This indicator was considered in relation to the value added from manufacturing for EU-27 countries [49,50]. Slovenia had a value added by manufacturing of 5.38 mil USD/capita, and 19% of its manufacturing enterprises used the IoT. Bulgaria had a value added by manufacturing of 2.64 mil USD/capita, but only 2% of its manufacturing enterprises used the IoT. Within the EU-27, Ireland had the largest value added by manufacturing, 26.05 mil USD/capita, but the IoT was implemented in only 7% of its manufacturing enterprises.

For I3, the best performer was Ireland, with an energy intensity of about 52 Kgoe/1000 EUR-PPS, while the worst was Malta, with an energy intensity of about 195 Kgoe/1000 EUR-PPS. Ireland was about four times less energy intensive than Malta (Kgoe/1000 EUR PPS = kilograms of oil equivalent per thousand euros in purchasing power standards).

For I4, the best performer was Denmark, with about 24 Mil EUR/100,000 inhabitants devoted to R&D, while the worst was Greece, with only 0.01 Mil EUR/100,000 inhabitants. Denmark invested 2400 times more than Greece in R&D.

For I5, the best performer was Sweden, with about 56% of its energy mix coming from RES, while the worst was Luxemburg, with only about 7%. Sweden had eight times more RES in its energy mix than Luxemburg.

For I6, the best performer was Finland, with approximately three patents/1000,000 inhabitants, while the worst were Bulgaria, Estonia, Greece, Croatia, Cyprus, Latvia, and Malta, with no patents.

Table 4 shows the results obtained by applying Phase 2 of this study, and shows the best and worst performers for each individual KPI.

Table 5 contains the results of the aggregation of KPIs after applying the last phases of this study, and allows the identification of three clusters.

Figure 3 contains the values of the ICEO, ranked from best to worst, and allows the identification of a variable gap between the leader and other member states, which ranges from 4% to 204%.

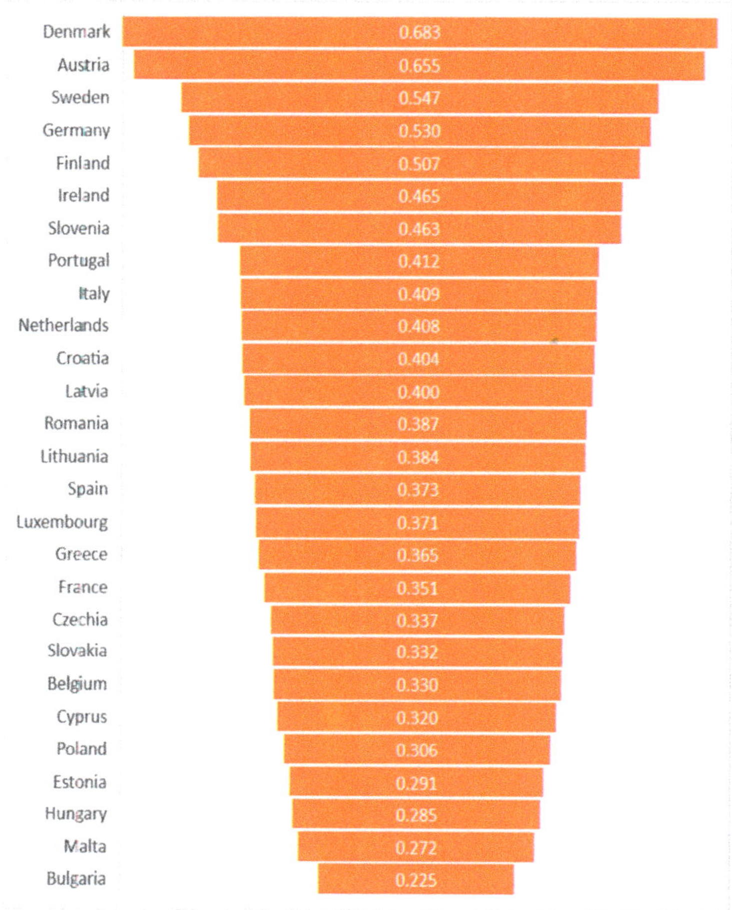


**Figure 3.** ICEO for EU-27 revealing the variability of both the ICEO and the gap behind the leading country.

## 4. Discussion

Unfortunately, the word still heavily relies on fossil fuels, which represented about 82% of primary energy consumption in 2021 [51], with only minor changes over the last decade. In the EU, the situation is slightly better, with only 70% of primary energy consumption coming from fossil fuels [51], but is still far from sustainable consumption. This leads to decisions related to production and consumption which are unstainable. Therefore, there is an increased need to develop, contextualize, and implement new multidimensional indicators to better track performance and progress. The ICEO may offer salient insights on the energy efficiency performance of the EU-27. The assessment is based on available indicators, as several possible KPIs have limited data availability.

The three factors of the CEO framework were considered equally important, as the best results will be obtained by considering managing consumption, targeting effectiveness, and creating opportunities in synergy. The three factors, regardless of the number of KPIs considered, must consider the reconciliation of the needs of society in terms of energy consumption with sustainable economic growth. In this way, the current model moves away from the dominant global economic paradigm and does not subordinate energy efficiency to economic growth at all costs. The study was restricted to the best few available

KPIs which avoided redundancy, overcrowding, and difficulty in application. The method identifies a single number (ICEO composite index) which captures the six included KPIs to synergically render an energy-efficiency ranking of each country belonging to the EU-27 space. The ICEO encapsulates the multi-dimensional concepts introduced by KPIs, different in meaning and units, allowing a ranking of EU-27 member states.

The values of the ICEO indicate that all EU-27 member states need to improve their energy systems. The average value of the ICEO was 0.400, less than half of the theoretical maximum of 1.00. Even the leading countries are far from the theoretical maximum value, except Denmark, which is the best performing country. The gap from the leader country, which is Denmark, varies from 4% (to the second country) to 204% (to the last country), showing that many countries lag behind and have an acute need for improvement.

Even if some countries are the leader for a given KPI, the overall result shows a balanced national approach is the key to an energy-efficient future. The present research identifies four clusters of countries:

Cluster I: Countries with significant performance, very actively involved on the road to an energy-efficient future (ICEO > 0.5). Five countries of the EU-27 are in this group, representing 19% of the EU-27 and being situated mostly in the northern part of EU-27: Denmark, Austria, Sweden, Germany, and Finland.

Cluster II: Countries with moderate performance (0.4 < ICEO < 0.5). Seven countries of the EU-27 are in this group, representing 25% of the EU-27: Ireland, Slovenia, Portugal, Italy, Netherlands, Croatia, and Latvia.

Cluster III: Countries with modest performance (0.3 < ICEO < 0.4). Eleven countries of the EU-27 are in this group, representing 41% of the EU-27, this cluster being the largest cluster: Romania, Lithuania, Spain, Luxembourg, Greece, France, Czechia, Slovakia, Belgium, Cyprus, and Poland.

Cluster IV: Countries with very modest performance (ICEO < 0.3). Four countries of the EU-27 are in this group, representing 14% of the EU-27: Estonia, Hungary, Malta, and Bulgaria.

From the best practice of the best performing countries, the slow performers should find suitable ways to accelerate transition to a more energy-efficient future. For instance, Denmark policy makers are declaring their country as "a laboratory for green solutions" [52], with a strong potential to share knowledge, especially in the field of wind energy and appropriate regulations. Learning from others means a faster pace of improvement, and is the best way to avoid making the same mistakes. Although the process of improvement takes time and cannot be achieved over night, it can be achieved faster. For instance, the countries belonging to Cluster IV should not wait as long as Denmark did to improve their energy system. Denmark started to take important steps towards its current status in 1970s, more than 50 years ago. This step-by-step process led to continuous growth over the last 40 years, with equal or lesser energy consumption. Every country must find the most appropriate mix of measures, but they must take into consideration the tackling of all possibilities in different degrees, appropriate to their economic structure, which range from the enhancement of energy intensity to the deployment of appropriate RES mix, and from increased investments to new regulations.

The main findings of this research are:

- In the EU-27, there are four clusters of countries, ranked based on their involvement in the transformation of their economies into more energy-efficient ones: active performers (Cluster I), moderate performers (Cluster II), modest performers (Cluster III), and slow performers (Cluster IV). Only 19% of EU-27 member states are in cluster I, showing a great need of improvement.
- Governance needs a deep reformation, as solutions must be applied systemically, at the needed pace and scale, by learning from the best practices already in effect in active performer countries.
- Energy-efficient measures must be tailored to regional and national contexts, bearing in mind best practices within countries with similar economic structure.

- Statistical monitoring related to energy savings and capital must be expanded.
- A European surveillance system of energy efficiency, which may share information and actions throughout EU-27, should be created.
- The effects of security crises on supply chains must be addressed, such as the recent invasion of Ukraine, by considering energy efficiency as a component of energy security.
- New technologies which are based on more efficient technologies, not linked to rebound effects that lead to more energy use, must be promoted.

The author acknowledges that energy efficiency may be evaluated in many ways, and that the performed analysis is subjective and not perfect. A holistic model for assessing energy efficiency, containing numerous KPIs, is difficult to consider, given the lack of availability of many data sets for all the countries of the EU-27, as well as to use, as many KPIs may overlap. The effectiveness of the proposed model rests in its straightforward application, based on a multidimensional framework which offers a quintessence of energy efficiency for all EU-27 member states, equally based on managing consumption, targeting effectiveness, and creating opportunities. The model best serves the question raised by the present study, which aims to offer important insights on the actual status of EU-27 member states on their road towards an energy-efficient future. Furthermore, the data were quantitatively expressed and retrieved from reliable international databases, without any overlap, while the methodology was based on data normalization and aggregation, and was unbiased and scientifically grounded.

The current study has limitations, most referring to the lack of available data. Moreover, this study does not apply to individual business, but judges as a whole the national systems. The proposed framework is suitable to analyse management and economic matters at aggregate level, but leaves out the complex performance tools used for businesses. This might constitute a direction for future research, as the topic needs further exploration. Energy performance tools are available for businesses, and they span from simple, based on energy monitoring, to complex, based on simulation [53]. However, these are mainly used in power generation sectors, as most of the manufacturing sector lacks data. Several real-time tools are available for businesses, such as continuous performance optimization [54] and energy management dashboards [55], but they are mainly process- or equipment-based tools. Moreover, artificial intelligence is also being considered for the self-optimization of plants, but such applications are not yet fully developed or available [53].

The conceptual foundation of the model reflects the author's personal view and crystallises the quintessence of the energy efficiency in three factors. The operational framework is not perfect, but allows valuable opinions which are context specific. The topic of energy efficiency is very complex and remains ontologically open. There are still research gaps to be filled, and a more in-depth analysis conducted by the actual providers of energy efficiency should be ascertained, as existing publications on SDG7 energy efficiency are mainly authored by international bodies [56,57]. The recent events of 2022 might boost awareness of energy efficiency, both for corporations and individual consumers, given the higher energy prices [56], but there are still major hurdles to be overcome, as much of the recorded decrease in energy consumption comes from a negative slowing of businesses. The topic of interwinding crises and their short- and long-term effects on energy efficiency may be a direction for future research, as this topic deserves additional exploration.

## 5. Conclusions

The present turbulent times, in which the world is still recovering from the COVID-19 pandemic, the invasion of Ukraine is ongoing, energy prices are volatile, and inflationary pressures are menacing political stability [57], it is up to governments to tackle the energy crisis as an opportunity or as a threat.

There is a great need for the transformation of the economies of the EU-27 to achieve the targets established by Agenda 2030 in terms of SDG7, given the current constraints. The EU-27 tends to be very differentiated and complex regarding the energy-efficiency future of each member state, but very similar in challenges: shortage of supply, climate change, increased prices, and dependency on imports. By spurring the best examples and avoiding the worst practices, learnt from the leading countries, the management of the whole energy system can be improved.

The findings of this research allow the ranking of countries from best to worst in terms of energy efficiency, and, by further studying the measures taken by leading countries, the identification of the best set of practices required by each country judged against its energy endowment; this knowledge may be a great tool in acquiring the targets which seem out of reach today. Moreover, the findings of this paper may launch new discussions among stakeholders in energy systems to find better ways to valorise energy efficiency.

**Funding:** This research received no external funding.

**Data Availability Statement:** Not applicable.

**Acknowledgments:** I am grateful to the perceptive reviewers for their insightful comments which have transformed the manuscript into a better one.

**Conflicts of Interest:** The author declares no conflict of interest.

## References

1. United Nations (UN). Transforming Our World: The 2030 Agenda for Sustainable Development. 2015. Available online: https://documents-dds-ny.un.org/doc/UNDOC/GEN/N15/291/89/PDF/N1529189.pdf?OpenElement (accessed on 2 November 2022).
2. International Energy Agency; International Renewable Energy Agency; United Nations Statistics Division; World Bank; World Health Organization. *Tracking SDG 7: The Energy Progress Report*; World Bank: Washington, DC, USA, 2022.
3. European Commission. *European Green Deal—Delivering on Our Targets*; Publications Office of the European Union: Luxembourg, 2021.
4. Rue du Can, S.; Letschert, V.; Agarwal, S.; Park, W.Y.; Kaggwa, U. Energy efficiency improves energy access affordability. *Energy Sustain. Dev.* **2022**, *70*, 560–568. [CrossRef]
5. European Parliament and of the European Council, Directive 2012/27/EU of the European Parliament and of the European Council on Energy Efficiency, 25 October 2012. Available online: https://eur-lex.europa.eu/LexUriServ/LexUriServ.do?uri=OJ:L:2012:315:0001:0056:en:PDF (accessed on 15 November 2022).
6. Islam, M.M.; Hasanuzzaman, M.D. Introduction to energy and sustainable development. In *Energy for Sustainable Development*; Hasanuzzaman, M.D., Rahim, N.A., Eds.; Academic Press: Cambridge, MA, USA, 2020; pp. 1–18. [CrossRef]
7. Chen, K.-H.; Cheng, J.-C.; Lee, J.-M.; Li, L.-Y.; Peng, S.-Y. Energy Efficiency: Indicator, Estimation, and a New Idea. *Sustainability* **2020**, *12*, 4944. [CrossRef]
8. Abu Bakar, N.N.; Hassan, M.Y.; Abdullah, H.; Rahman, H.A.; Abdullah, P.; Hussin, F.; Bandi, M. Energy efficiency index as an indicator for measuring building energy performance: A review. *Renew. Sustain. Energy Rev.* **2015**, *44*, 1–11. [CrossRef]
9. European Parliament. *Proposal for a Directive of the European Parliament and of the Council on Energy Efficiency, Brussels, 14.7.2021 COM(2021) 558 Final*; Publications Office of the European Union: Luxembourg, 2021.
10. European Commission. *Sustainable Development in the European Union—Monitoring Report on Progress Towards the SDGs in an EU Context*; Publications Office of the European Union: Luxembourg, 2022.
11. Council of European Energy Regulators, 2nd CEER Report on Power Losses, C19-EQS-101-03, 23 March 2020. Available online: https://www.ceer.eu/documents/104400/-/-/fd4178b4-ed00-6d06-5f4b-8b87d630b060 (accessed on 23 November 2022).
12. Dean, N. Gains and losses across Europe. *Nat. Energy* **2019**, *4*, 92. [CrossRef]
13. International Energy Agency, Policies Database. 2022. Available online: https://www.iea.org/policies?topic[]=Energy+Efficiency (accessed on 28 November 2022).
14. Eurostat, Energy Dashboard. 2022. Available online: https://ec.europa.eu/eurostat/cache/infographs/energy_dashboard/endash.html (accessed on 28 November 2022).
15. Wuppertal Institute, EMEEES Project—Measuring and Reporting Energy Savings for the Energy Services Directive—How It Can Be Done. 2009. Available online: https://wupperinst.org/en/p/wi/p/s/pd/155 (accessed on 8 November 2022).
16. *ISO 50001*; ISO Organization: Geneva, Switzerland, 2018. Available online: https://www.iso.org/iso-50001-energy-management.html (accessed on 8 November 2022).
17. Rossiter, A.; Jones, B. (Eds.) *Energy Management and Efficiency for the Process Industries*; John Wiley & Sons, Inc.: Hoboken, NJ, USA, 2015; p. xi.
18. Johansson, M.; Thollander, P. A review of barriers to and driving forces for improved energy efficiency in Swedish industry—Recommendations for successful in-house energy management. *Renew. Sustain. Energy Rev.* **2018**, *82*, 618–628. [CrossRef]
19. Hasan, A.S.M.M.; Rokonuzzaman, M.; Tuhin, R.A.; Salimullah, S.M.; Ullah, M.; Sakib, T.H.; Thollander, P. Drivers and Barriers to Industrial Energy Efficiency in Textile Industries of Bangladesh. *Energies* **2019**, *12*, 1775. [CrossRef]

20. Bagaini, A.; Colelli, F.; Croci, E.; Molteni, T. Assessing the relevance of barriers to energy efficiency implementation in the building and transport sectors in eight European countries. *Electr. J.* **2020**, *33*, 106820. [CrossRef]
21. Cai, X.; Wang, W.; Rao, A.; Rahim, S.; Zhao, X. Regional Sustainable Development and Spatial Effects from the Perspective of Renewable Energy. *Front. Environ. Sci.* **2022**, *10*, 859523. [CrossRef]
22. Economist Intelligence, Energy Outlook 2023, Surviving the Energy Crisis. 2022. Available online: https://www.eiu.com/. (accessed on 30 November 2022).
23. International Energy Agency. *Tracking Public Investment in Energy Technology Research—A Roadmap*; IEA: Paris, France, 2021.
24. European Commission. *Communication from the Commission to the European Parliament, the European Council, the Council, the European Economic and Social Committee and the Committee of the Regions REPowerEU Plan*; COM(2022) 230 Final; European Commission: Brussels, Belgium, 2022.
25. European Commission. *Communication from the Commission to the European Parliament, the Council, the European Economic and Social Committee and the Committee of the Regions EU 'Save Energy'*; COM(2022) 240 Final; European Commission: Brussels, Belgium, 2022.
26. European Commission. *Report from the Commission to the European Parliament, the Council, the European Economic and Social Committee and the Committee of the Regions State of the Energy Union 2022*; COM(2022) 547 Final; European Commission: Brussels, Belgium, 2022.
27. Holechek, J.L.; Geli, H.M.E.; Sawalhah, M.N.; Valdez, R. A Global, Assessment: Can Renewable Energy Replace Fossil Fuels by 2050? *Sustainability* **2022**, *14*, 4792. [CrossRef]
28. Future Earth; The Earth League; WCRP. *10 New Insights in Climate Science 2022*; WCRP: Stockholm, Sweden, 2022.
29. Gil, D.; Ferrández, A.; Mora-Mora, H.; Peral, J. Internet of Things: A Review of Surveys Based on Context Aware Intelligent Services. *Sensors* **2016**, *16*, 1069. [CrossRef] [PubMed]
30. Ramallo-González, A.P.; Bardaki, C.; Kotsopoulos, D.; Tomat, V.; González Vidal, A.; Fernandez Ruiz, P.J.; Skarmeta Gómez, A. Reducing Energy Consumption in the Workplace via IoT-Allowed Behavioural Change Interventions. *Buildings* **2022**, *12*, 708. [CrossRef]
31. Al-Obaidi, K.M.; Hossain, M.; Alduais, N.A.M.; Al-Duais, H.S.; Omrany, H.; Ghaffarianhoseini, A. A Review of Using IoT for Energy Efficient Buildings and Cities: A Built Environment Perspective. *Energies* **2022**, *15*, 5991. [CrossRef]
32. Reuter, M.; Patel, M.; Eichhammer, W.; Lapillonne, B.; Pollier, K. A comprehensive indicator set for measuring multiple benefits of energy efficiency. *Energy Policy* **2020**, *139*, 111284. Available online: https://www.sciencedirect.com/science/article/pii/S0301421520300434 (accessed on 11 January 2023). [CrossRef]
33. Subramanian, S.; Bastian, H.; Hoffmeister, A.; Jennings, B.; Tolentino, C.; Vaidyanathan, S.; Nadel, S. *International Energy Efficiency Scorecard*; American, Council for an Energy-Efficient Economy: Washington, DC, USA, 2022. Available online: www.aceee.org/research-report/i2201 (accessed on 3 November 2022).
34. Lapillonne, B. *Monitoring EU Energy Efficiency First Principle and Policy Implementation—ODYSSEE MURE*; Enerdata: Grenoble, France, 2020.
35. Singh, H.V.; Bocca, R.; Gomez, P.; Dahlke, S.; Bazilian, M. The energy transitions index: An analytic framework for understanding the evolving global energy system. *Energy Strategy Rev.* **2019**, *26*, 100382. [CrossRef]
36. Bruni, G.; De Santis, A.; Herce, C.; Leto, L.; Martini, C.; Martini, F.; Salvio, M.; Tocchetti, F.A.; Toro, C. From Energy Audit to Energy Performance Indicators (EnPI): A Methodology to Characterize Productive Sectors. The Italian Cement Industry Case Study. *Energies* **2021**, *14*, 8436. [CrossRef]
37. Purvis, B.; Mao, Y.; Robinson, D. Three pillars of sustainability: In search of conceptual origins. *Sustain. Sci.* **2019**, *14*, 681–695. [CrossRef]
38. Nolting, L.; Steiger, S.; Praktiknjo, A. Assessing the validity of European labels for energy efficiency of heat pumps. *J. Build. Eng.* **2018**, *18*, 476–486. [CrossRef]
39. Marshall, E.; Steinberger, J.; Dupont, V.; Foxon, T. Combining energy efficiency measure approaches and occupancy patterns in building modelling in the UK residential context. *Energy Build.* **2016**, *111*, 98–108. [CrossRef]
40. Momete, D.C. Measuring Renewable Energy Development in the Eastern Bloc of the European Union. *Energies* **2017**, *10*, 2120. [CrossRef]
41. Momete, D.C. A unified framework for assessing the readiness of European Union economies to migrate to a circular modelling. *Sci. Total Environ.* **2020**, *718*, 137375. [CrossRef] [PubMed]
42. Eurostat. Online Data Base, Primary Energy Consumption, Code [SDG_07_10], Last Update 11 of July 2022. Available online: https://ec.europa.eu/eurostat/cache/metadata/en/sdg_07_10_esmsip2.htm (accessed on 5 November 2022).
43. Eurostat. Online Data Base, Final Energy Consumption [SDG_07_11] Last Update 11 of July 2022. Available online: https://ec.europa.eu/eurostat/cache/metadata/en/sdg_07_11_esmsip2.htm (accessed on 5 November 2022).
44. Eurostat. Online Data Base, Energy Intensity, Code [NRG_IND_EI], Last Update 11 April 2022. Available online: https://ec.europa.eu/eurostat/cache/metadata/en/nrg_ind_ei_esmsip2.htm (accessed on 30 October 2022).
45. Eurostat. Online Data Base, Internet of Things Code [ISOC_EB_IOT$DEFAULTVIEW], Last Update 17 of March 2022. Available online: https://ec.europa.eu/eurostat/cache/metadata/en/isoc_e_esms.htm (accessed on 30 October 2022).
46. SETIS Research and Innovation Data. Available online: https://setis.ec.europa.eu/publications/setis-research-and-innovation-data_en (accessed on 30 October 2022).

47. Eurostat. Online Data Base, Share of Renewable Energy in Gross Final Energy Consumption by Sector Code [SDG_07_40], Last Update 19 April 2022. Available online: https://ec.europa.eu/eurostat/cache/metadata/en/sdg_07_40_esmsip2.htm (accessed on 30 October 2022).
48. Eurostat. Online Data Base, Patents Related to Recycling and Secondary Raw Material, Code [CEI_CIE020], Last Update 5 July 20222. Available online: https://ec.europa.eu/eurostat/cache/metadata/en/cei_cie020_esmsip2.htm (accessed on 30 October 2022).
49. Eurostat. Online Data Base, Population on 1 January, Code [TPS00001], Last Update 29 October 2022. Available online: https://ec.europa.eu/eurostat/databrowser/view/tps00001/default/table (accessed on 30 October 2022).
50. World Bank Online Data Base, Manufacturing, Value Added (Current US$). Available online: https://data.worldbank.org/indicator/NV.IND.MANF.CD (accessed on 22 December 2022).
51. British Petroleum. *BP Statistical Review of World Energy 2022*; BP: London, UK, 2020. Available online: https://www.bp.com/en/global/corporate/energy-economics/statistical-review-of-world-energy.html (accessed on 2 December 2022).
52. Official Website of Denmark. Available online: https://denmark.dk/ (accessed on 30 November 2022).
53. Australian Government and Advisian, Energy Efficiency Tools Review, Australia. 2020. Available online: https://www.energy.gov.au/sites/default/files/Energy%20Efficiency%20Tools%20Review_1.pdf (accessed on 30 November 2022).
54. Siemens, Continuous Performance Optimization. Available online: https://www.siemens-energy.com/global/en/offerings/services/digital-services/continuous-performance-optimization.html (accessed on 30 October 2022).
55. Antsolutions, Energy Management Dashboard. Available online: https://antsolutions.eu/products/energy-management-system-ems/ (accessed on 30 November 2022).
56. International Energy Agency. *Energy Efficiency 2022*; International Energy Agency: Paris, France, 2022.
57. Cozzi, L.; Gould, T. (Eds.) *World Energy Outlook 2022*; International Energy Agency: Paris, France, 2022.

*energies*

*Article*

# "Is Energy That Different from Labor?" Similarity in Determinants of Intensity for Auto Assembly Plants

**Amir Abolhassani [1,*], Gale Boyd [1], Majid Jaridi [2], Bhaskaran Gopalakrishnan [2] and James Harner [3]**

1 Social Science Research Institute & Department of Economics, Duke University, Durham, NC 27708, USA
2 Industrial and Management Systems Engineering Department, West Virginia University, Morgantown, WV 26506, USA
3 Department of Statistics, West Virginia University, Morgantown, WV 26506, USA
* Correspondence: amir.abolhassani@duke.edu

**Abstract:** This paper addresses the question "Is energy that different from labor?" from the perspective of efficiency. It presents a novel statistical analysis for the auto assembly industry in North America to examine the determinants of relative energy intensity, and contrasts this with a similar analysis of the determinants of another important factor of production, labor intensity. The data used combine two non-public sources of data previously used to separately study key performance indicators (KPIs) for energy and labor intensity. The study found these two KPIs are statistically correlated (the correlation coefficient is 0.67) and the relationship is one-to-one. The paper identifies 11 factors that may influence both energy and labor intensity KPIs. The study then contrasts which of the empirical factors the two KPIs' share and how they differ. Two novel statistical methods, Huber estimators and Multiple M-estimators, combined with regularized algorithms, are identified as the preferred methods for robust statistical models to estimate energy intensity. Based on our analysis, the underlying determinants of energy efficiency and labor productivity are quite similar. This implies that strategies to improve energy may have spillover benefits to labor, and vice versa. The study shows vehicle variety, car model types, and launch of a new vehicle penalize both energy and labor intensity, while flexible manufacturing, production volume, and year of production improve both energy and labor intensity. In addition, the study found that the plants that produce small cars are more energy-efficient and productive compared to plants that produce large vehicles. Moreover, in a given functional unit, i.e., on a per-unit basis, Japanese plants are more energy-efficient and productive compared to American plants. Plant managers can use the proposed data-driven approach to make the right decisions about the energy efficiency targets and improve plants' energy efficiency up to 38% using hybrid regression methods, mathematical modeling, plants' resources, and constraints.

**Keywords:** energy-efficient manufacturing; productivity; unit energy intensity; automotive industry

## 1. Introduction

The climate crisis and its consequence for life on the planet require a global awareness about green manufacturing, particularly energy efficiency. In recent years, many governments, the general public, and manufacturing companies have focused on efficiency as a primary way to both save money and reduce emissions. Nowadays, companies are not only concerned about their financial business performance but also about the impact of their manufacturing operations on the environment [1–4].

In 2013, industrial companies' energy consumption was 22% of total U.S. energy consumption and ranked second after transportation [5]. It was estimated that the U.S. automotive sector's energy use was over 800 trillion BTU per year [6]. However, if the contribution of the substantial supply chain is included, the total energy consumption related to the car manufacturing industry will be considerably higher. A typical car manufacturing facility consumes different sources of energy, of which electricity and natural gas are the two most predominant ones. Electricity cost is higher between the two and represents 62% of the total energy costs in a car manufacturing facility [1,2].

The automotive industry has witnessed several challenges that pressure this sector to change its business model and reprioritize its long-term strategy. Some of those include cost pressure, lower sales volume, electric vehicles (EVs), autonomous vehicles (AVs), and carbon footprint reductions that are forced by national and international regulators [7,8]. During the last few decades, many governments and industrial sectors have shown steady progress in improving energy efficiency; however, the question of how the industrial sectors can accelerate this journey remain controversial [9]. Some automakers have already started to reduce their carbon footprint by heavily investing in the renewable energy resources. They want to hedge against possible future energy price rise and also meet their social responsibility commitment. Some have invested as much as USD 1 billion to build farms that can use wind, solar panels, landfill gas, or biogas to produce electricity and have plants with zero $CO_2$ emission by 2040 [10,11].

Although considerable research has been devoted to aggregate manufacturing energy efficiency, less attention has been paid to the factors that reduce energy consumption at a plant level. This paper employs a non-public plant-level dataset along with robust statistical models to estimate plant-level energy efficiency and labor productivity. The first contribution of this paper is the merging of two non-public datasets, previously used to study energy and labor separately. The second contribution is using a systematic approach to identify the factors that enhance both plant's energy efficiency and labor productivity (The terms "efficiency" and "productivity" are used interchangeably in this paper. The notions of efficiency and productivity are simply "two sides of the same coin" but arise from the different approaches in the literature on energy efficiency and labor productivity.) Using the unique dataset and novel statistical methods, the paper examines synergies between energy intensity and labor intensity, as well as differences.

The remainder of the paper is organized as follows. Related studies, research gaps, and industrial needs are reviewed and discussed in Section 2. Research methodology is presented in Section 3. Data preparation and empirical analysis are examined in Section 4. The conclusion and implications of this study are discussed in Section 5.

## 2. Literature Review

This section provides an overview of the recent studies on productivity and energy efficiency and attempts to identify the most common productivity KPI, energy efficiency KPI, and their underlying determinants. It categorizes the current research into two main categories, empirical and theoretical research, and builds upon their limitations. The empirical research includes engineering assessments and case studies, and theoretical research includes surveys and analytical and review papers. Section 2 is organized in three subsections: productivity studies, energy efficiency studies, and the link between the two.

### *2.1. Productivity KPI in This Study*

The most expensive element in an OEM (Original Equipment Manufacturer) is the supply chain, while the second is labor cost [12]. Labor contributes 12 to 15 percent of the total production costs for a car manufacturer [13]. Nearly all major automakers are collaborating with the Oliver Wyman Institute to help in assessing their manufacturing plants' labor productivity. Hours per vehicle (HPV) is the most commonly utilized labor productivity KPI. HPV measures the amount of all direct and indirect labor hours (salaried and hourly employees) needed to assemble a vehicle in body, paint, and trim shops; for more details, readers are referred to Harbour reports [14]. This study uses HPV as the most common productivity KPI to measure the labor intensity and investigates its relationship with the energy efficiency KPI, energy intensity.

### 2.2. Studies That Influence Energy Efficiency

The authors found that there are two main categories of energy efficiency in the literature, empirical research and theoretical research [15]. The authors summarize the existing studies and review their limitations in the coming subsections.

#### 2.2.1. Empirical Research

##### Energy Efficiency through Assessments and Exergy Analysis

Engineering assessment studies are mostly of saving energy through developing smart systems or improving the current technology at a manufacturing plant. Galitsky et al. (2008) conducted a comprehensive study about different energy-saving opportunities within U.S. automakers. The authors categorized energy efficiency opportunities either in the area of utility systems or process systems. These utility systems include heat and steam distribution, motors, lighting, material handling, heating ventilation and air conditioning (HVAC), and general. Process-related systems include stamping, welding, and painting [16]. These studies provide recommendations to improve energy efficiency based on technical reports and case studies, including energy efficiency experiences of Asian and European, as well as American car makers. They found paint processes, HVAC, and lighting were the top three consumers of energy, respectively. Kluczek (2019) examined 12 U.S. manufacturing facilities and showed energy intensity could be improved through life cycle assessments (LCAs) and slacks-based measure (SBM). He proposed a comprehensive and sustainable methodology to improve energy intensity in the production systems [17]. Another common traditional practice is exergy analysis, which improves energy efficiency through the second law of thermodynamics [18]. The exergy analysis is a practical method that focus on energy conservation and energy distribution process systems.

The purpose of this research is different and complementary to the studies reported above. This research aims to define the factors to enhance energy intensity via an advanced statistical approach, without focusing on utility systems and process energy efficiency recommendations.

##### Enhancing Energy Efficiency through Case Studies and Focus Groups

The most common types of qualitative research are case studies and focus groups. Feng and Mears (2016) studied the energy consumption in a car manufacturing plant that utilized electricity and natural and landfill gases. The authors calculated the energy consumption percentages for different departments such as body, paint, and assembly. Additionally, they estimated the monthly energy consumption of the studied plant based on heating and cooling degree days [19]. May et al. (2015, 2016) carried out semi-structured interviews of six European companies. They proposed a tailored and adapted seven-step method to develop energy key performance indicators (e-KPIs) [20,21]. The authors proposed some conceptual frameworks to measure the e-KPIs but were not able to analytically examine and validate the model.

Generally, case studies and focus groups study one or few manufacturing facilities, which potentially limits their generalizability. The current study uses additional variables to measure energy efficiency within a plant.

#### 2.2.2. Theoretical Research

There are several theoretical research approaches such as survey, simulation, and analytical, which will be discussed in the next subsections.

##### Enhancing Energy Efficiency through Survey

Damert and Baumgartner (2018) studied a sample of 116 global car companies in America, Europe, Japan, and South Korea. Damert and Baumgartner found that OEMs' climate strategies are significantly different from suppliers and OEMs had stringent climate policies. Moreover, they found that the main contributors of an organization's climate policy were regional affiliations, its home country's institutional environment, and the size of the company [22]. Hence, the current study considers plants' ownership as a factor that may capture various OEMs' policy or organizations' culture and leadership style, which may directly/indirectly affect energy efficiency.

Enhancing Energy Efficiency through Analytical Approach

Boyd (2014, 2017) developed an energy performance indicator (EPI) to enable identification of the "best in class" energy performance for an industry. The EPI assigns a specific energy performance score on a 1–100 scale to a plant [23,24]. The author used energy usage, annual production, plant utilization rate, HDD and CDD for the plant location, and the wheelbase of the largest vehicle produced in that plant. Oh and Hildreth (2014) carried out a study using stochastic frontier and data envelopment analysis methods to measure GM plants' effectiveness of energy-saving initiatives. They used Spearman's rank-order coefficient and log likelihood tests to check the stochastic frontier analysis model. They proposed a benchmarking process to estimate the energy intensity per vehicle by using variables such as HDD and CDD, plant utilization, and the wheelbase of the largest vehicle produced in the plant [25]. Jeon et al. (2015) estimated the combined energy consumption of both natural gas and electricity of US SMEs in the metal sector. They estimated the annual energy consumption of a manufacturing facility by using various variables such as number of employees, annual production level and hours, and plant area [26]. Alsaleh and Abdul-Rahim (2018, 2019) used the generalized method of moments (GMM) in multiple studies to review the impact of numerous factors on bioenergy intensity in European Countries [27–29]. They found that the biggest contributors to bioenergy intensity were growth of gross domestic product and government investment. The intent of our research is different, and the focus of this study is on manufacturing facilities' energy efficiency rather than improving the energy efficiency of countries and governmental policies. Williams et al. (2020) utilized a network analysis to design a green supply chain. They considered a network of customers and store locations as a supply chain where the customers could drive to or from stores, and retailers supplied the materials to stores from a central warehouse [30]. They identified the number of stores in the supply chain by minimizing the greenhouse gas (GHG) emission. They found doubling the fuel efficiency of customers' vehicles reduced the supply chain emissions by 46% compared to the baseline scenario.

The research presented in this paper considers additional variables to current studies in the literature. Moreover, several robust methods, including shrinkage and hybrid methods, are utilized to develop robust statistical models that are not sensitive to unusual data points. Furthermore, this study compares the underlying determinants of energy efficiency and labor productivity. The variables and analytical approach are thoroughly discussed in Sections 3 and 4.

Enhancing Energy Efficiency through Review Papers

Artificial intelligence, digital twins, and digital technology can be used throughout the vehicles lifecycle management to enhance sustainability and quality of the products [31]. Schulze et al. (2016) reviewed 44 scientific papers to define the major elements of energy management. The authors proposed a conceptual framework that has five major key elements: strategy, implementation, controlling, organization, and culture [32]. May et al. (2017) also reviewed 365 articles to examine the main lines of the energy management. They proposed a framework with six essential areas: strategic paradigms, Information and Communication Technologies (ICTs), drivers and barriers, supporting tools and methods, production–performance trade-off, and production process paradigms [33].

Various researchers conducted review studies to define implemented energy efficiency methods in high-energy-consuming industries. Moreover, the authors reviewed four energy efficiency evaluation methods such as stochastic frontier, DEA, benchmarking comparison, and exergy analysis [18,34–36]. The intent of this research is different compared to the stated review studies, and it examines a longitudinal study at the plant level. Additionally, the current study proposes a structured methodology that could measure and estimate the plants' energy efficiency.

2.2.3. Association between Energy Efficiency and Labor Productivity KPIs

Different types of studies like surveys, case studies, and statistical analysis have examined the correlation between the plants' energy efficiency and productivity. Some researchers studied the correlation between green manufacturing and lean manufacturing, which is almost equivalent to energy efficiency and productivity. This subsection reviews the literature related to productivity and energy efficiency association.

King and Lenox (2001) conducted a comprehensive study on 17,499 U.S. manufacturing facilities over five years, 1991–1996. The authors used a Probit model and two-stage least squares methods in their analysis. They found that organizations who implemented the quality management system standard, ISO 9001, were more likely to also adopt the environmental management standard, ISO 14000 [37].

Boyd and Pang (2000) carried out a research study on 171 U.S. glass manufacturing facilities to consider the correlation between energy efficiency and productivity. Productivity was measured by using a frontier production function approach to define the gap between a hypothetical plant and a benchmark of best practice. They found that a 1% productivity improvement resulted in more than 1% energy enhancement for the flat glass manufacturers; however, the result was not significant for the container glass manufacturers [4].

Worrell et al. (2003) examined the association between energy efficiency and productivity enhancement measures over 70 industrial plants in the U.S. They discovered that energy efficiency investments could boost the overall productivity in the industry [38].

Bergmiller and McCright (2009) surveyed 47 American companies to study the synergy between lean and green programs. They found green manufacturing drivers could result in significant achievements in lean practices, specifically cost performance enhancement. Moreover, the number of people dedicated to waste projects could be reduced because of implementing the projects that benefit both green and lean projects [3]. Inman and Green (2018) carried out a similar study on 182 manufacturing managers in the U.S. and found similar results [39].

Sobral et al. (2013) conducted an in-depth case study utilizing in-person interviews and direct observations in a multinational car maker. They interviewed different people like the plant's environmental technicians, lean initiators, and managers in both production and environmental departments. They found lean manufacturing practices can create sustainable environmental benefits [1]. In a similar study, Cherrafi et al. (2017) showed companies could save 20–40% of their energy consumption by implementing lean and green manufacturing simultaneously [40]

Diaz-Elsayed et al. (2013) developed an approach to incorporate the lean and green strategies in a production system. They carried out a discrete event simulation study on a part supplier in the automotive industry. They noticed that implementing the combination of lean and green practices reduced the production lead time and production cost by 10.8% [41].

Chiarini (2014) conducted a case study on five European automotive suppliers to investigate the relation between a few specific lean practices and green manufacturing. He used cellular manufacturing, value stream mapping, single minute exchange of die, total productive maintenance, and 5S as lean practices. He found all studied variables, except for the single minute exchange of die, can improve energy efficiency and create sustainable manufacturing processes [42].

Herrmann et al. (2008) conducted a simulation study at a European organization to investigate the synergy between lean and green manufacturing. They used five lean practices as a representative of lean manufacturing. The lean practices used were pull-system standardization, total quality management, Andon line, and continuous flow practices [43]. The scholars found that standardization practices resulted in lower energy consumption, but other lean practices increased energy consumption per product unit.

Boyd and Curtis (2014) studied the linkages between management practices and both energy efficiency and workforce productivity in the U.S. [44]. Unlike Bloom's study in the UK, they found no link between good management and energy efficiency [45].

Several studies found green supply chain management practices could lead to a better organizational performance in different areas such as environmental, economic, performance, and social [46–48].

Wei and Liu (2017) studied the global rebound outcomes on energy use and emissions resulting from energy efficiency improvement. The scholars used a Computable General Equilibrium (CGE) model to evaluate the energy efficiency improvement compared to a business-as-usual scenario. They estimated a very large rebound outcome on energy use and related emissions (70% and 90%, respectively) in 2040 at the global level. Capital and labor supply are the most important factors that can play a critical role in developing an effective energy supply in the long term. Rath et al. (2019) examined the relation of energy consumption and total factor productivity (TFP) for 36 countries. They found that the use of fossil fuels reduces the growth of TFP, whereas the use of renewable energy increases the growth of TFP [49,50]. The study examined by Chen et al. found that the relation between total labor costs and energy intensity was negative for 22 emerging economies [51].

Many successful case studies have been carried out to explore the productivity and energy efficiency association. However, case studies may potentially limit the generalizability. Additionally, survey questions are usually designed so that a variety of respondents can answer them. Therefore, survey studies may not be as substantial as studies utilizing empirical studies and secondary analysis. Furthermore, there are some contradictions between various researchers about the relation between productivity and energy efficiency, or lean manufacturing and green manufacturing. Hence, the authors of the research presented here decided to examine the relation between energy efficiency and labor productivity using a secondary analysis which is more generalizable compared to case studies and surveys.

Table 1 presents a summary of available research and the existing research limitations.

**Table 1.** Existing literature research methods and limitations.

| Research Areas | Research Approach | Research Methods | References | Limitations |
|---|---|---|---|---|
| Enhancing energy efficiency | Empirical | Engineering assessment | Hildreth (2014), [25]<br>Thollander et al. (2007), [32]<br>Galitsky et al. (2008), [16]<br>Li and Tao (2017), [18] | They were concentrated on the energy assessment only, and no mathematical or statistical modeling that could assist other manufacturing plants were provided. |
| | | Focus group and case study | Feng & Mears (2016), [19]<br>May et al. (2016), [21]<br>May et al. (2015), [20] | In general, a case study potentially limits its generalizability. |
| | Technical | Survey | Damert and Baumgartner (2018), [22] | The study only considered the effect of ownership on energy efficiency. |
| | | Analytical | Boyd (2014, 2017). [23,44]<br>Jeon et al. (2015), [26]<br>Oh and Hildreth (2014), [25]<br>Alsaleh and Abdul-Rahim (2018, 2019), [27–29] | These studies used very few variables to evaluate the energy efficiency. This paper includes several new variables to measure the energy efficiency. |
| | | Review papers | Li and Tao (2017), [18]<br>May et al. (2017), [33]<br>Mardani et al. (2017), [34]<br>Schulze et al. (2016), [32]<br>Prabhu et al. (2015), [35]<br>Bhattacharya et al. (2015), [36] | They presented a very high level of information and developed a conceptual framework which may not be practical at the plants' operational level. |

**Table 1.** *Cont.*

| Research Areas | Research Approach | Research Methods | References | Limitations |
|---|---|---|---|---|
| Energy efficiency and productivity association | Empirical | Case study (Interview, simulation, value stream mapping) | Cherrafi et al. (2017), [40] Chiarini (2014), [42] Sobral et al. (2013), [1] Diaz-Elsayed et al. (2013), [41] Herrmann et al. (2008), [43] | Since they were case studies, they potentially limit their generalizability. |
| | Technical | Survey | Bergmiller and McCright (2009), [3] King and Lenox (2001), [37] Inman and Green (2018), [39] | In general, a survey's validity in scientific research could be an issue. |
| | | Statistical | Boyd and Curtis (2014), [44] Bloom et al. (2010), [45] Boyd and Pang (2000), [4] | There are contrary results regarding the energy efficiency and productivity association. |

This study is built upon several research gaps identified in Table 2 that have been found during the literature review. These gaps and research questions can be addressed by investigating the relationship between energy efficiency and productivity, defining the underlying determinants of energy efficiency, and comparing those energy efficiency determinants to the productivity determinants. Table 2 highlights the research gaps and questions, industry needs, areas of focus, and degree of contribution.

**Table 2.** The research gaps and industrial needs.

| Research Gaps/Objectives | Research Questions | Industry Needs | Areas of Focus | Degree of Contribution |
|---|---|---|---|---|
| Gap 1 There are contrary results in the literature regarding the association between energy efficiency and productivity KPIs | RQ1. Is there a significant correlation between energy efficiency and productivity? | The plant managers may have better insight into the association between productivity and energy efficiency KPIs | Energy efficiency and productivity association | Modest contribution |
| Gap 2 There are few studies with statistical analysis approach defining the impact of important variables and factors on energy efficiency KPI | RQ2. What are the most important factors that can improve the energy efficiency? | The auto plants can use the developed analytical method to improve the energy efficiency | Energy efficiency | Decisive contribution |
| Gap 3 No study has compared the underlying determinants of energy efficiency and productivity KPIs | RQ3. What are the common determinants that can improve both productivity and energy efficiency KPIs at the same time? | The plant managers may have a better comprehension of the factors that can improve productivity and energy efficiency | Energy efficiency and productivity association | Decisive contribution |

The research gaps and questions are discussed in detail in the following sections.

## 3. Research Methodology

This paper uses the Unit Energy Intensity (UEI) to represent the energy efficiency KPI. The UEI is the plant's annual energy consumption divided by the total number of vehicles produced in a year. The plant's location may have a significant contribution to the amount of annual energy consumption. Therefore, the effect of HDD and CDD must be considered and UEI needs to be adjusted accordingly. Galitsky et al. (2008) found that

11–20% of the energy consumption might be contributed to comfort heating and cooling. In this study, it was assumed that at most 14% of electricity consumption is used for comfort cooling if a plant is cooled, and at most 17% of natural gas consumption is used for comfort heating. The adjusted UEIs for electricity and natural gas consumption could be defined using adjusted HDD and CDD as follows:

$$UEI_{Elec-adj} = \frac{\text{Annual electricity usage}}{\text{Annual production volume}} \times (1 - 0.14 \times CDDadj)$$

$$UEI_{NG-adj} = \frac{\text{Annual natural gas usage}}{\text{Annual production volume}} \times (1 - 0.17 \times HDDadj)$$

where

$$CDD_{adj} = \frac{CDD - Min(CDD)}{Max(CDD) - Min(CDD)}$$

If a plant is not cooled, the $CDD_{adj} = 0$. Min(CDD) and Max(CDD) are calculated from all the plants' locations during the study period.

$$HDD_{adj} = \frac{HDD - Min(HDD)}{Max(HDD) - Min(HDD)}$$

Min(HDD) and Max(HDD) are calculated from all the plants' location during the study period. All the participants' plants are heated during the winter months.

Hence, the $UEI_{Elec\text{-}adj}$ and $UEI_{NG\text{-}adj}$ could be calculated as described (This assumes that the average energy share is linear in the range of HDD and CDD. Further investigation confirmed that $UEI_{Elec\text{-}adj}$ and $UEI_{NG\text{-}adj}$ are linear in the range of HDD and CDD). After a literature review, interviewing experts who had considerable expertise in industry practice and academia, and practical consideration, 11 factors have been identified for analysis and their impact on UEIs. Some factors have never been utilized in the context of UEIs and some variables' definitions have been changed by authors. These factors are: vehicle segment, launching a new vehicle, year, car assembly and capacity (CAC) utilization, flexible manufacturing (FM), total number of chassis configurations and body styles (vehicle variety), number of annual working days (AWD), number of models produced, annual production volume (APV), whether or not the plant is cooled (AC), and plant ownership. The summary of these factors, including their references, are shown in Table 3. Moreover, 11 hypotheses were posed in Table 3 for further investigation.

**Table 3.** The important factors/variables that might impact on UEIs.

| Factor/Variable | Comments/References | Definition | Research Hypotheses |
|---|---|---|---|
| Productivity KPI | [14] | HPV represents the productivity KPI. HPV is calculated by dividing the total working hours of certain hourly and salaried employees by the total number of vehicles produced in a calendar year. These working hours are measured during all stages of vehicle production and processing including body, painting, and final assembly. | NA |
| Energy efficiency KPI | Not used in the literature | In our study, UEI is the energy efficiency KPI and calculated as plant's annual energy consumption divided by the total number of vehicles produced in a year. | NA |
| Vehicle segment | Not used in the literature to estimate UEIs | Vehicle segment is defined as cars' class. The authors used the same vehicle segment categories which were developed to measure HPV [52] | H1. Vehicle segment has a significant impact on the UEIs |

**Table 3.** *Cont.*

| Factor/Variable | Comments/References | Definition | Research Hypotheses |
| --- | --- | --- | --- |
| CAC Utilization | [14,23–25,33] | CAC utilization is the total number of cars manufactured per design capacity line in December of the data year. CAC can be calculated as:<br>Annual capacity = 235 days per year × 16 h per day × Capacity line rate (except where otherwise stated, capacity line rate is based on December)<br>CAC = (Total number of cars manufactured/Annual Capacity) × 100 | H2. UEIs are correlated with CAC Utilization |
| Vehicle variety | Not used in the literature to estimate UEIs | It considers the total number of car body types and chassis configurations.<br>- Number of car body types: It describes the number of variations in body types such as convertible, passenger, cargo van, wagon, 2-door, 3-door, 4-door, and 5-door [14].<br>- Number of chassis: It describes the number of variations in chassis configuration such as rear or front wheel drive [14]. | H3. UEIs are correlated with the vehicle variety |
| Number of models | Not used in the literature to estimate UEIs | It describes how many various car nameplates are manufactured in a plant. | H4. UEIs are correlated with the number of models |
| APV | [23–26] | APV represents the total number of cars manufactured by the end of the year (i.e., in December), except where otherwise stated. | H5. UEIs are correlated with APV |
| Plant ownership | [20,22]<br>Not used in the literature to estimate UEIs | Plant ownership may drive differences in terms of plants' working culture, management style, technologies, work regime, and regulations. Damert and Baumgartner (2018) and May et al. (2015) indirectly considered plant ownership as a factor that might impact UEIs. | H6. UEI variation for American and Japanese is significantly different |
| Flexible manufacturing (FM) | Not used in the literature to estimate UEIs | Flexible manufacturing involves using standard procedures, common locator points, shared skid systems, etc. It assists manufacturers in producing better cars at cheaper price while preserving quality. FM used in this study incorporates the flexibility of the equipment, volume, mix, and utilization, which reflect the effects of market variations and economies of scale.<br>The FM utilized in this study can be determined as follows:<br>$FM = Ln\,(F_{Eq} \times F_V \times F_M \times F_U)$<br>where<br>$F_{Eq}$ = The equipment flexibility, described as:<br>$\frac{\text{No. of body styles}}{\text{No. of body lines}} \times \frac{\text{No. of body styles}}{\text{No. of paint lines}} \times \frac{\text{Vehicle variety}}{\text{No. of assembly lines}}$<br>$F_V$ = The production volume, referred to as volume flexibility<br>$F_M$ = The various number of nameplates that are produced in a factory, referred to as mix flexibility<br>$F_U$ = The CAC, referred to as utilization flexibility | H7. UEIs are correlated with flexible manufacturing |
| Air conditioning (AC) | Not used in the literature to estimate UEIs | AC is a binary variable and set equal to one when a plant is cooled during the summer months, otherwise it is set to zero. This variable is used for the energy source of electricity only. | H8. UEIs are correlated with AC |
| Vehicle launch | Not used in the literature to estimate UEIs | A great amount of effort is put forth by engineering, supply chain, production, logistics, and quality teams to overcome the challenges of a new product launch. The study shows that the plants encountered some difficulties keeping their UEIs while introducing new vehicles. | H9. UEIs are correlated with launching a new vehicle |
| AWD | Not used in the literature to estimate UEIs | AWD describes the number of scheduled working days for the data year. The AWD does not account for planned holidays and vacations (such as summer and New Year shutdowns). Throughout the study period, the AWD for all brand names was essentially uniform and fell between 230 and 245 days. | H10. UEIs are correlated with AWD |
| Year of production | Not used in the literature to estimate UEIs | Year is the fiscal year that the data are provided for | H11. UEIs are correlated with year of production |

Table 3 indicates that, prior to this research, eight factors/variables have not been utilized in the literature to estimate the energy efficiency KPI, UEIs. In this study, the factors that are used to estimate the UEIs include: year of UEI measurement, vehicle segment, CAC utilization, new vehicle launch, vehicle variety, number of car models produced, APV, FM, AWD, whether or not the plant is cooled, and plant ownership.

The research steps, data preparation, and empirical analysis conducted in this research are shown in Figure 1. Figure 1a summarizes the research steps and Figure 1b summarizes the data preparation and analysis. The research questions and hypotheses are discussed in detail in Section 4.

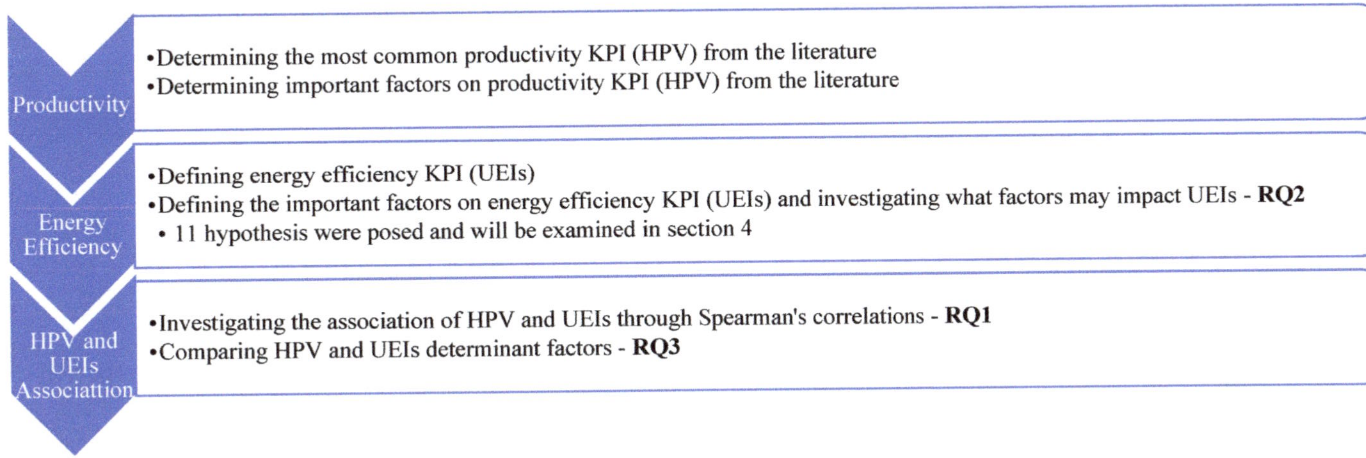


(a)

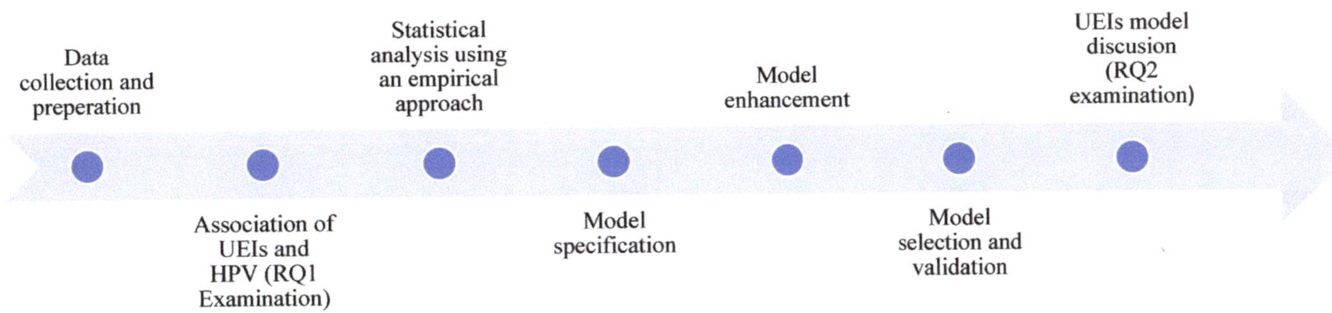


(b)

**Figure 1.** Synopsis of research steps (**a**) and data preparation and empirical analysis (**b**).

## 4. Data Preparation and Empirical Analysis

Data collection and preparation, the association between UEIs and HPV, the empirical approach to statistical analysis, model specification and refinement, and model selection and validation are reviewed in this section.

### *4.1. Data Collection and Preparation*

The data analyzed in this study covered a 6-year period, 1999–2005 (excluding 2002), for the North American Automakers involved in the development of the Energy Star EPI (The Energy Star data were collected in two waves (as described in Boyd 2014). During the first round, two plants provided data for 2001 that caused a data gap in this year.) and Harbour Institute's assessments. The authors used a secondary analysis on a combined dataset based on the Energy Star and Harbor reports. The combined dataset has been used to identify and develop UEIs and important determinants on UEIs and HPV.

Once the data from Energy Star and the Harbour reports were collected, edit checks were performed, and the $UEI_{Elec\text{-}adj}$ and $UEI_{NG\text{-}adj}$ were plotted versus all variables, other than nominal variables, to discover any typographical error or extreme unusual data points. Some outliers from the original dataset were found, and it was decided not to remove them, but rather to minimize their impact by applying a method to down-weight them. The number of plants of two ownerships (American: Ford and GM, Japanese: Honda, Nummi, and Toyota) used in this research is presented in Table 4.

**Table 4.** Frequency of data for different car manufacturers in the study period.

| Ownership | 1999 | 2000 | 2001 | 2003 | 2004 | 2005 | Total * |
|---|---|---|---|---|---|---|---|
| American | 25 | 23 | 0 | 20 | 21 | 21 | 110 |
| Japanese | 2 | 2 | 2 | 4 | 4 | 8 | 22 |
| Total | 27 | 25 | 2 | 24 | 25 | 29 | 132 |

* One observation that has a huge discrepancy between Energy Star and Harbour reports dataset was deleted from the analysis.

The descriptive statistics for $UEI_{Elec\text{-}adj}$ and $UEI_{NG\text{-}adj}$ over the study period are presented in Appendix A, Table A1. The $UEI_{Elec\text{-}adj}$ and $UEI_{NG\text{-}adj}$ variation for different ownership categories in the study period is illustrated in Figures 2 and 3. These figures indicate that Japanese ownership had a lower value of $UEI_{Elec\text{-}adj}$ and $UEI_{NG\text{-}adj}$ during the study period. However, the actual comparisons may slightly change because of utilizing the UEIs which were not adjusted for the climate.

Abolhassani and Jaridi (2016) found that vehicle segmentation was an important factor/variable in estimating the HPV. In order to reduce the vehicle segmentation's variation, they recommended to reduce the number of vehicle segments from 14 to 3 [52]. The same vehicle class grouping is used in this study. It was decided to combine group C with group B to enhance the significance of the statistical analysis for all the groups and create a balanced sample (group C had six records only). As shown in Table 5, two vehicle segments are used in this study.

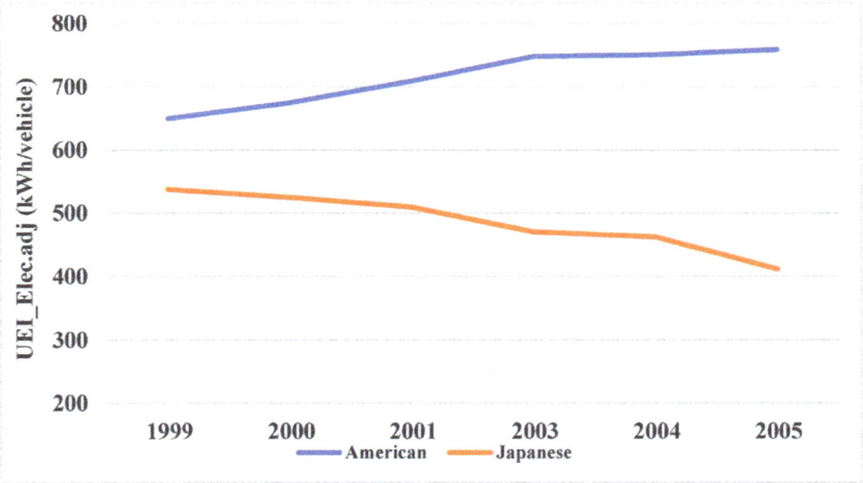


**Figure 2.** $UEI_{Elec\text{-}adj}$ for different ownership in the study period (kWh/vehicle).

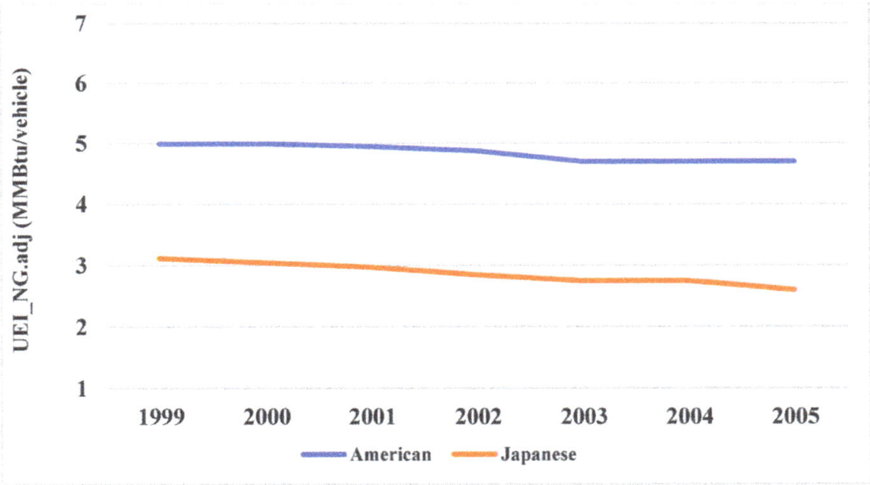


**Figure 3.** $UEI_{NG\text{-}adj}$ for different ownership in the study period (MMBtu/vehicle, 1 MMBtu ≈ 1055 MJ).

**Table 5.** Different vehicle segmentation.

| Segment | Used Grouping | Frequency of Data |
|---|---|---|
| Subcompact, Midsize, Midsize SUV, Small Pickup | A | 64 |
| Compact, Small SUV, Sports Car, Luxury, Large, Minivan, Large Van, Full-Size SUV, Full-Size Pickup, Medium Duty | B | 68 |
| Total | | 132 |

The mean of HPV varies from 20 to 25 and from 25 to 35 for groups A and B, respectively. Subcompact, midsize, midsize SUV, and small pickup served as group A and the remaining segments as group B. Different vehicle segmentations are shown in Table 5.

The minimum sample size for a statistical model must be greater than 2R + 25, where R is the number of regressors [53]. In the following statistical model, 12 regressors, including the intercept, are used to estimate the response. Hence, R = 12 and the sample size must be greater than 49. Therefore, the number of data points used to develop the UEIs' statistical models is sufficient for both car groups.

### *4.2. Association of UEIs and HPV and RQ1 Examination*

In this subsection, the first research question

RQ1: Is there a significant correlation between energy efficiency and productivity?

is examined and discussed.

A correlation analysis is used to estimate the association between UEIs and HPV. There are several methods of correlation analysis, e.g., Pearson, Spearman's rank order, and Kendall's tau. The Pearson correlation is used when the data follow a normal distribution. However, if normality does not hold, the non-parametric Spearman's rank-order or Kendall's tau correlations are more appropriate [54].

The Shapiro and Wilk test can be used to check whether the data are normally distributed. If the *p*-value obtained in this test is less than or equal to a desired $\alpha$, in this study 0.05, the null hypothesis (normality) is rejected. Failing to reject normality indicates that there is not enough evidence to suggest the null hypothesis is false at the 1-$\alpha$ confidence level [55]. The Shapiro and Wilk test and related *p*-value are shown in Table 6 for HPV, $UEI_{Elec\text{-}adj}$, and $UEI_{NG\text{-}adj}$.

**Table 6.** Shapiro and Wilk test for HPV, $UEI_{Elec\text{-}adj}$, and $UEI_{NG\text{-}adj}$.

| | **Shapiro and Wilk Test Statistic** | ***p*-Value** |
|---|---|---|
| HPV | 0.82142 | $2.01 \times 10^{-11}$ |
| $UEI_{Elec\text{-}adj}$ | 0.78284 | $9.35 \times 10^{-13}$ |
| $UEI_{NG\text{-}adj}$ | 0.79855 | $3.106 \times 10^{-12}$ |

Table 6 indicates that HPV, $UEI_{Elec\text{-}adj}$, and $UEI_{NG\text{-}adj}$ do not follow the normal distribution. Hence, the non-parametric correlation tests, like Spearman or Kendall, must be used. Usually, Kendall is applied when there are tied values of the response variables; however, Spearman is applied for series of data without tied values. Because there are no tied values in this dataset, the authors use Spearman's correlation test. The Spearman's correlation test is carried out between HPV and UEIs. The null hypothesis shows that there is no monotonic correlation between HPV and UEIs, versus the alternative hypothesis, which indicates the existence of a monotonic correlation. The Spearman's correlation test between the HPV and UEIs is given in Table 7.

**Table 7.** Spearman's correlation test statistics for the HPV and UEIs.

| | **Variables** | **HPV** | **$UEI_{Elec\text{-}adj}$** | **$UEI_{Elec\text{-}adj}$** |
|---|---|---|---|---|
| All plants (n = 132) | HPV | 1.0000 | | |
| | $UEI_{Elec\text{-}adj}$ | 0.6764<br>0.0000 | 1.0000 | |
| | $UEI_{NG\text{-}adj}$ | 0.6777<br>0.0000 | 0.7937<br>0.0000 | 1.0000 |

Table 7 indicates that there is a statistically significant positive correlation between HPV and UEIs; hence, RQ1 is addressed. Moreover, the correlation coefficients between HPV and $UEI_{Elec\text{-}adj}$ and $UEI_{NG\text{-}adj}$ are almost the same. The Spearman's correlation between HPV and $UEI_{Elec\text{-}adj}$ is 0.6764, meaning there is a significantly large and positive relationship between these two variables.

The results of the Spearman's correlation analysis for different variables and various scenarios are shown in Table 8.

**Table 8.** Spearman's correlations and the corresponding *p*-values by ownership, American manufacturers, and utilizing air conditioning.

| | **Variables** | **HPV** | $\mathbf{UEI_{Elec\text{-}adj}}$ | $\mathbf{UEI_{Elec\text{-}adj}}$ |
|---|---|---|---|---|
| | | Plants' ownership | | |
| American (n = 110) | HPV | 1.0000 | | |
| | $UEI_{Elec\text{-}adj}$ | 0.6698<br>0.0000 | 1.0000 | |
| | $UEI_{NG\text{-}adj}$ | 0.6810<br>0.0000 | 0.7557<br>0.0000 | 1.0000 |
| Japanese (n = 22) | HPV | 1.0000 | | |
| | $UEI_{Elec\text{-}adj}$ | −0.1914<br>0.3935 | 1.0000 | |
| | $UEI_{NG\text{-}adj}$ | −0.4225<br>0.0501 | 0.4654<br>0.0291 | 1.0000 |
| | | Vehicle launch | | |
| No vehicle launch (n = 106) | HPV | 1.0000 | | |
| | $UEI_{Elec\text{-}adj}$ | 0.5882<br>0.0000 | 1.0000 | |
| | $UEI_{NG\text{-}adj}$ | 0.5902<br>0.0000 | 0.7449<br>0.0000 | 1.0000 |
| Vehicle launch (n = 26) | HPV | 1.0000 | | |
| | $UEI_{Elec\text{-}adj}$ | 0.7483<br>0.0000 | 1.0000 | |
| | $UEI_{NG\text{-}adj}$ | 0.8509<br>0.0000 | 0.8995<br>0.0291 | 1.0000 |
| | | Air conditioning | | |
| AC (n = 48) | HPV | 1.0000 | | |
| | $UEI_{Elec\text{-}adj}$ | 0.6836<br>0.0000 | 1.0000 | |
| | $UEI_{NG\text{-}adj}$ | 0.5659<br>0.0000 | 0.7557<br>0.0000 | 1.0000 |
| No AC (n = 84) | HPV | 1.0000 | | |
| | $UEI_{Elec\text{-}adj}$ | 0.7544<br>0.000 | 1.0000 | |
| | $UEI_{NG\text{-}adj}$ | 0.7549<br>0.0000 | 0.7436<br>0.0000 | 1.0000 |
| | | American plants | | |
| GM (n = 63) | HPV | 1.0000 | | |
| | $UEI_{Elec\text{-}adj}$ | 0.5552<br>0.0000 | 1.0000 | |
| | $UEI_{NG\text{-}adj}$ | 0.6145<br>0.0000 | 0.6058<br>0.0000 | 1.0000 |
| Ford (n = 47) | HPV | 1.0000 | | |
| | $UEI_{Elec\text{-}adj}$ | 0.8000<br>0.000 | 1.0000 | |
| | $UEI_{NG\text{-}adj}$ | 0.7649<br>0.0000 | 0.7239<br>0.0000 | 1.0000 |

The findings from Table 8 are listed as follows (at the 0.05 significance level), The impacts of plant size on UEIs are provided on Table A2, Appendix A:

- The Spearman's correlation between HPV and UEIs is statistically significant for American companies; however, it is not significant for Japanese plants. Intuitively, the correlation coefficients between HPV and UEIs are different for plant characteristics/ownership. The authors used Fisher transformation to statistically investigate the difference between the plants' ownership. The reader should refer to Appendix A and Table A3 for a detailed methodology. However, it must be emphasized that the Fisher transformation on the rank transformation, Spearman rank, is an approximation and the reader must be cautious in their interpretation. Furthermore, the reader must be cautious about multiple testing issues (i.e., the Bonferroni corrections must be used to compensate for the multiple group effects).
- The Spearman's correlations between HPV and UEIs while launching a new vehicle are higher when there is no new vehicle launch.
- The Spearman's correlations between HPV and UEIs for the plants that are not cooled are higher compared to plants that are cooled.
- The Spearman's correlations between HPV and UEIs for Ford are higher than GM.

### 4.3. Empirical Approach to the Statistical Analysis

A multiple linear regression model, MLRM, is selected where UEIs are considered as the response, with 11 and 10 regressors for electricity and natural gas, respectively. Multiple transformations of each original continuous regressor, including cube, square, inverse cube, square root, logarithm, and reciprocal, were examined to determine the most appropriate transformation for each continuous regressor. Initially, 11 regressors (or their transformed regressors) were selected for more in-depth analysis.

The UEI statistical model formulation is as follows:

$$\widehat{UEI}_{Elec/NG-adj} = \hat{\beta}_0 + \sum_{i=1}^{P} \hat{\beta}_i \, F(x_i) \quad (1)$$

where

$\widehat{UEI}_{Elec/NG-adj}$ = The estimated value of adjusted unit energy intensity; the units are kWh or MMBtu (1 MMBtu ≈ 1055 MJ) for electricity and natural gas, respectively;

P = 11 (the number of regressors);

F(x) = Identity function or indicator of nominal variable or real transformation (real transformation can include cube, square, inverse cube, square root, logarithm, and reciprocal of any of the genuine continuous variables);

$X_1$ = APV, no units;

$X_2$ = FM, no units;

$X_3$ = AWD, days;

$X_4$ = Year, no units;

$X_5$ = CAC utilization, no units;

$X_6$ = Vehicle variety, no units;

$X_7$ = Number of car models, no units;

$X_8$ = New vehicle launch, no units;

$X_9$ = Vehicle segment (A and B), no units;

$X_{10}$ = AC (Air conditioning status; it is used when the energy source is electricity), no units;

$X_{11}$ = Plant ownership (American and Japanese), no units.

### 4.4. Model Specification

The Akaike Information Criteria (AIC) are commonly used with the goal of identifying the most appropriate combination of regressors to adequately describe the interrelationship between the regressors and the response, UEIs. Furthermore, AIC can be used for the statistical modeling and variable selection [56].

To enhance the predictive power of the statistical model, some transformations were performed. The preliminary statistical model was developed using stepwise regression with AIC for all 11 independent variables (there would be 10 variables when the energy source is natural gas), in which CAC and APV are transformed. The significant regressors (at the 10% significance level) and resulting coefficient estimates are demonstrated in Tables 9 and 10 for $UEI_{Elec\text{-}adj}$ and $UEI_{NG\text{-}adj}$, respectively.

**Table 9.** Initial multiple linear regression for the $UEI_{Elec\text{-}adj}$.

| Regressor/Variable | Coefficient Estimates | Std. Error | t Value | Pr(>\|t\|) | |
|---|---|---|---|---|---|
| Intercept | 47,839.52 | 14,327.59 | 3.339 | 0.001118 | ** |
| AWD | 3.0655 | 0.8523 | 3.597 | 0.000468 | *** |
| Vehicle Variety | 11.128 | 4.5255 | 2.459 | 0.015348 | * |
| No. of Models | 68.2555 | 15.1413 | 4.508 | $1.52 \times 10^{-5}$ | *** |
| Year | −20.533 | 7.1228 | −2.883 | 0.004667 | ** |
| ln(APV) | −461.527 | 61.8735 | −7.459 | $1.45 \times 10^{-11}$ | *** |
| ln(CAC) | −308.371 | 81.6402 | −3.777 | 0.000247 | *** |
| Plant Ownership (Japanese *) | −99.0017 | 52.0287 | −1.903 | 0.059439 | . |
| Vehicle Launch | 78.9496 | 41.6989 | 1.893 | 0.060703 | . |
| Segment B ** | 99.2191 | 36.0382 | 2.753 | 0.006812 | ** |
| AC | 121.8007 | 36.5979 | 3.328 | 0.001159 | ** |
| Significance Level Codes: '***' 0.001, '**' 0.01, '*' 0.05, '.' 0.1, ' ' 1 | | | | | |
| Residual standard error: 176.2 on 121 degrees of freedom | | | | | |
| Multiple R-squared: 0.7931, Adjusted R-squared: 0.776 | | | | | |
| F-statistic: 46.39 on 10 and 121 DF, $p$-value: $<2.2 \times 10^{-16}$ | | | | | |
| AIC = 1375.82 | | | | | |

* Japanese plants are contrasted with American plants. ** Segment B is contrasted with segment A.

**Table 10.** Initial multiple linear regression for the $UEI_{NG\text{-}adj}$.

| Regressor/Variable | Coefficient Estimates | Std. Error | t Value | Pr(>\|t\|) | |
|---|---|---|---|---|---|
| (Intercept) | 578.5916 | 139.2828 | 4.154 | $6.03 \times 10^{-5}$ | *** |
| Vehicle Variety | 0.36758 | 0.07104 | 5.174 | $8.93 \times 10^{-7}$ | *** |
| FM | −0.32597 | 0.15538 | −2.098 | 0.037941 | * |
| Year | −0.25658 | 0.06932 | −3.701 | 0.000321 | *** |
| ln(APV) | −4.1223 | 0.61132 | −6.743 | $5.24 \times 10^{-10}$ | *** |
| ln(CAC) | −2.33311 | 0.78209 | −2.983 | 0.003436 | ** |
| Plant Ownership (Japanese *) | −0.80878 | 0.46742 | −1.73 | 0.086064 | . |
| Segment.B ** | 0.56456 | 0.3485 | 1.62 | 0.107781 | |
| Significance Level Codes: '***' 0.001, '**' 0.01, '*' 0.05, '.' 0.1, ' ' 1 | | | | | |
| Residual standard error: 1.761 on 124 degrees of freedom | | | | | |
| Multiple R-squared: 0.7272, Adjusted R-squared: 0.7118 | | | | | |
| F-statistic: 47.23 on 8 and 124 DF, $p$-value: $<2.20 \times 10^{-16}$ | | | | | |
| AIC = 157.2 | | | | | |

* Japanese plants are contrasted with American plants. ** Segment B is contrasted with segment A.

Here are the results of the $UEI_{Elec\text{-}adj}$ statistical model:

- The F-test ($p$-value $< 2.20 \times 10^{-16}$) determines that the null hypothesis could not be statistically rejected and the $UEI_{Elec\text{-}adj}$ can be estimated based on a set of regressors or their transformation.
- Out of 11 regressors, ten variables, including AWD, air conditioning, vehicle variety, vehicle segment, number of models, plant ownership, year of survey, APV, and CAC, are statistically significant at the 10% significance level.
- The adjusted r-squared indicates that the set of ten selected regressors explains 77% of the variation in $UEI_{Elec\text{-}adj}$.

Here are the results of the $UEI_{NG\text{-}adj}$ statistical model:

- The F-test ($p$-value $< 2.20 \times 10^{-16}$) determines that the null hypothesis could not be statistically rejected and the $UEI_{NG\text{-}adj}$ can be estimated based on a set of regressors or their transformation.
- Out of 10 (AC was excluded because the energy source is natural gas) regressors, six including vehicle variety, FM, year of survey, APV, CAC, and plant ownership are statistically significant at the 10% significance level.
- The adjusted R-squared indicates that the set of six selected regressors explains 71% of the variation in $UEI_{Ng\text{-}adj}$.

A thorough examination of the residuals shows that all underlying assumptions of the regression analysis are met. In order to explore the possible endogeneity and simultaneity issues, the authors performed the serial correlation for the dependent variables (UEIs) over the study period. For the white noise series, it is expected that each autocorrelation be close to zero. Furthermore, 95% of the spikes in the autocorrelation function, ACF, lies within $\pm\frac{2}{\sqrt{T}}$, where $T$ is the length of the time series. It is common to plot the control bands on a graph of the ACF. If there are one or more large spikes outside these control bands, or if more than 5% of spikes are outside these control bands, then the series is possibly not white noise [57]. The result of the serial correlation for the UEIs and different combination of Japanese/American, vehicle segment A/B, and whether or not the plant was launching a new vehicle indicate that there is no serial correlation over the study period and the autocorrelation plots show independent white noise. In addition, the authors use four control variables (plant ownership, vehicle segment, launching a new vehicle, and plant air conditioning status) to avoid any possible endogeneity [58].

### 4.5. Model Enhancement

Some unusual and high-leverage data points are identified when building the multiple linear regression model in the previous subsection. These data points were not data entry errors, nor did they come from a different population than most of the data. There was therefore no compelling reason to exclude these plants from the analysis. It was decided not to exclude these plants from the analysis and to include all of them; subsequently, robust regression is used as a good way to down-weight unusual data points [59,60].

There are several statistical modeling techniques to improve the performance of a statistical model. These methods include all robust methods (M-Estimation, MM-Estimation, SMDM-estimators), shrinkage methods (lasso, ridge, and elastic net), and the combination of robust and shrinkage methods. Here is the list of statistical methods used and developed:

1. M-Estimation: M-estimators are a broad class of robust regression methods that also include the maximum likelihood estimators [61]. Huber and Tukey bisquare weight functions are the most common M-estimators [62].
2. MM-Estimation: MM estimator (or Multiple M-estimator) is a special form of M-estimation. It is a combination of high breakdown-point and efficient estimation, S estimation [61,63].
3. SMDM-Estimation: The scholars suggested additional steps to enhance the MM-estimation. They used an adaptive method to re-estimate the regression parameters and utilize them for the initial estimate in multiple M-estimation [64,65].

4. Shrinkage and Regularized Regression: Shrinkage and regularized methods are typically used to optimize the bias–variance trade-off. The three most popular shrinkage estimators include ridge, lasso, and elastic net regressions and are determined as follows:

$$\text{Ridge regression}: \sum_{i=1}^{n}\left(y_i - \beta_0 - \sum_{j=1}^{p}\beta_j x_{ij}\right)^2 + \lambda\sum_{j=1}^{p}\beta_j^2 = \text{RSS} + \lambda\sum_{j=1}^{p}\beta_j^2 \quad (2)$$

$$\text{Lasso regression}: \sum_{i=1}^{n}\left(y_i - \beta_0 - \sum_{j=1}^{p}\beta_j x_{ij}\right)^2 + \lambda\sum_{j=1}^{p}\left|\beta_j\right| = \text{RSS} + \lambda\sum_{j=1}^{p}\left|\beta_j\right| \quad (3)$$

$$\begin{aligned}\text{Elastic net}: &\sum_{i=1}^{n}\left(y_i - \beta_0 - \sum_{j=1}^{p}\beta_j x_{ij}\right)^2 + \lambda\sum_{j=1}^{p}\left((1-\alpha)\beta_j^2 + \alpha\left|\beta_j\right|\right) \\ &= \text{RSS} + \lambda\sum_{j=1}^{p}\left((1-\alpha)\beta_j^2 + \alpha\left|\beta_j\right|\right)\end{aligned} \quad (4)$$

The tuning parameter, $\lambda$, is identified via an iterative process and minimizing the cross-validation mean squared prediction error. Then, the shrinkage and regularized regression coefficient estimates are determined using the obtained tuning parameter and refitting the regression model. The regularized methods usually do not hold any assumptions and converge faster than ordinary least squares [66]. Hence, there is no need to perform any further residual analysis, and the coefficient estimates could be considered as a proper value of the actual coefficients.

Elastic net regression is a hybrid method that combines both ridge and lasso. The term $(1-\alpha)\beta_j^2$ coerces the correlated coefficient estimates to be averaged, whereas the term $\alpha\left|\beta_j\right|$ promotes a sparse solution in the coefficients of the averaged features [67]. The $\alpha$, hyper-parameter, varies between 0 and 1 and controls the $L_1$ and $L_2$ penalization (0 is ridge and 1 is lasso).

The authors did not optimize the objective function, Equation (4), in two-dimensional space, for both $\alpha$ and $\lambda$. They optimized it in one dimension, $\lambda$ only, and used equal spaces of $\alpha$ ($\alpha = 0, 0.25, 0.5, 0.75, 1$) to check if there are significant differences on the model's performance when $\alpha$ varies.

5. **Hybrid Regression Methods:** There are some outliers in the pooled dataset; therefore, robust methods can be used to down-weight these unusual observations. Furthermore, shrinkage methods can be used to shrink the coefficient estimates and take care of any possible collinearity. Therefore, the combination of the robust and shrinkage methods was used. Consequently, the weight of each observation (i.e., each plant) was calculated, and a matrix of weights was formed in which the diagonal is the weights of plants for a particular robust method and the off-diagonal matrix is zero. The matrix of weights is then used in the shrinkage methods to reduce any possible collinearity. Utilizing the weights of each plant and considering the shrinkage methods, Equations (2)–(4) may be rewritten as:

$$\text{Ridge regression}: \min_\beta (Y - X\beta)^T W_m (Y - X\beta) + \lambda\sum_{j=1}^{p}\beta_j^2 \quad (5)$$

$$\text{Lasso regression}: \min_\beta (Y - X\beta)^T W_m (Y - X\beta) + \lambda\sum_{j=1}^{p}\left|\beta_j\right| \quad (6)$$

$$\text{Elastic net}: \min_\beta (Y - X\beta)^T W_m (Y - X\beta) + \lambda\sum_{j=1}^{p}\left((1-\alpha)\beta_j^2 + \alpha\left|\beta_j\right|\right) \quad (7)$$

where

$W_m$ is an n x n matrix where n = 132 (n is the sample size and equals the number of data points). $W_m$ is a diagonal matrix where the elements off the main diagonal are all zero, and the diagonal is the data points' weights (i.e., $W_{im}$, i = 1, 2, . . . , n). The weights also depend on m, which indexes the robust method type (i.e., M-estimators (Huber and Tukey bisquare), MM-estimator, and SMDM-estimator).

The $W_m$ matrix for the Huber method (m = 1) can be represented as follows:

$$W_m = \begin{bmatrix} w_{1,1} & 0 \cdots & 0 \\ \vdots & \ddots & \vdots \\ 0 & \cdots & w_{132,1} \end{bmatrix}$$

Model selection and validation are discussed in the next subsection.

### *4.6. Model Selection and Validation*

Researchers commonly use cross-validation as an important step to validate the final model. There are two main objectives which cross-validation can help: addressing any possible overfitting and calculating the mean squared error, MSE.

In K-fold cross-validation, the dataset is divided into K samples with equal sizes known as K folds. Among K equal folds, K − 1 folds are utilized to train the model and the model performance is evaluated on the last fold. The leave-one-out cross-validation, LOOCV, is another common approach that can be used in model validation. In LOOCV, the dataset is split into K folds where K equals the number of data points (K = n). However, LOOCV can be computationally expensive to use if the sample size is large [68].

According to Kohavi (1995), ten equal folds (K = 10) produces better results than other numbers of folds such as LOOCV, five, or twenty folds. With the goal of achieving the highest level of accuracy, ten folds were applied to evaluate the model performance.

The developed statistical model can be used to estimate $\hat{Y}_{ij}$, where i represents the observation and j represents the fold. The error can be calculated by determining the difference between the response and its estimated value, as $Y_{ij} - \hat{Y}_{ij}$, for all the K folds. Therefore, the cross-validation error, $CV_{(k)}$, can be calculated as,

$$CV_{(k)} = \sum_{i=1}^{k} \frac{n_i \times MSE_i}{n}$$

($n_i$ is the number of data points in fold $i$).
Where

$$MSE_i = \sum_{i\in k^{th}fold} \frac{(Y_i - \hat{Y_i})^2}{n_i}$$

Tables 11 and 12 summarize all developed regression models for electricity and natural gas. These include the base model (MLRM with data transformation) and improved models: all robust regressions, shrinkage and regularized regressions, and the combination of robust and shrinkage techniques.

**Table 11.** Performance comparison of the developed statistical models for electricity (using 10-fold cross-validation).

| Energy Source | Regression Method | Estimation Type | Adjusted R-Squared | Residual Standard Error (RSD) | Cross-Validation with 10 Folds (MSE) |
|---|---|---|---|---|---|
| $\hat{UEI}_{Elec-adj}$ | MLRM | MLRM with transformation | 0.77 | 176.2 | 188.50 |
| | Robust Regression | M-estimator (Huber) | 0.79 | 121 | 186 |
| | | M-estimator (Tukey bisquare) | 0.62 | 86.4 | 221 |
| | | MM-estimator | 0.78 | 101 | 222 |
| | | SMDM-estimator | 0.79 | 152 | 236 |

**Table 11.** *Cont.*

| Energy Source | Regression Method | Estimation Type | Adjusted R-Squared | Residual Standard Error (RSD) | Cross-Validation with 10 Folds (MSE) |
|---|---|---|---|---|---|
| | Shrinkage Methods | Ridge Regression, α = 0 | 0.79 | | 187.22 |
| | | Lasso regression, α = 1 | 0.80 | | 188.47 |
| | | Elastic net, α = 0.50 | 0.79 | | 187.22 |
| | Hybrid (Robust and Shrinkage) Methods | Lasso with M-estimator (Huber) | 0.83 | | 153 |
| | | Lasso with M-estimator (Tukey bisquare) | 0.80 | | 88.85 |
| | | Lasso with MM-estimator | 0.81 | | 103.51 |

**Table 12.** Performance comparison of the developed statistical models for natural gas (using 10-fold cross-validation).

| Energy Source | Regression Method | Estimation Type | Adjusted R-Squared | Residual Standard Error (RSD) | Cross-Validation with 10 Folds (MSE) |
|---|---|---|---|---|---|
| $\hat{UEI}_{NG-adj}$ | MLRM | MLRM with transformation | 0.71 | 1.76 | 1.93 |
| | Robust Regression | M-estimator (Huber) | 0.72 | 1.58 | 2.03 |
| | | M-estimator (Tukey bisquare) | 0.70 | 1.51 | 2.09 |
| | | MM-estimator | 0.67 | 1.40 | 2.05 |
| | | SMDM-estimator | 0.65 | 1.60 | 2.22 |
| | Shrinkage Methods | Ridge Regression, α = 0 | 0.72 | | 1.94 |
| | | Lasso regression, α = 1 | 0.73 | | 1.93 |
| | | Elastic net, α = 0.50 | 0.73 | | 1.94 |
| | Hybrid (Robust and Shrinkage) Methods | Lasso with M-estimator (Huber) | 0.73 | | 1.63 |
| | | Lasso with M-estimator (Tukey bisquare) | 0.70 | | 1.56 |
| | | Lasso with MM-estimator | 0.73 | | 1.39 |
| | | SMDM-estimator | 0.73 | | 1.54 |

Here are the findings from Tables 11 and 12.

- Multiple linear regression has higher residual standard error (RSD) compared with robust regressions.
- Multiple linear regression has lower adjusted R-squared compared with regularized regressions and robust regressions.
- The hybrid methods (combination of regularization and robust) have almost the highest adjusted R-squared and lowest cross-validation errors compared with other methods.

Table 11 shows that the adjusted R-squared improved from 0.77 (base Elec model) to 0.83 (Huber lasso) and from 0.71 (base NG model) to 0.73 (MM lasso); however, the cross-validation errors were reduced from 188.49 to 153 and from 1.93 to 1.39 for electricity and natural gas, respectively. Moreover, a careful examination of Tables 11 and 12 confirms the existence of some unusual observations in our study, lower residual standard error for robust regressions compared to MLRM, and the rationale that hybrid methods have better regression fitting performance.

On top of the statistical models' performance, the regressors' coefficient signs were carefully examined for different statistical models that were presented in Tables 11 and 12. Almost all the estimated coefficients in different statistical models had the same sign except for the vehicle launch variable. For electricity energy source, the sign of this variable was positive when multiple linear regression or lasso with M-estimator (Huber) was used. It means that while launching a new vehicle the electricity consumption per vehicle increases. The authors investigated this fact through data visualization and conducting a paired sample *t*-test. The results showed that while launching a new vehicle, $\hat{UEI}_{Elec-adj}$ increased. Therefore, the authors decided to use the lasso with M-estimator (Huber) and lasso with MM-estimator, which show better statistical performance compared to MLRM for electricity and natural gas, respectively.

### *4.7. UEI Model Discussion and RQ2 Examination*

In this subsection, the second research question

RQ2: What are the most important factors that can improve the energy efficiency?

and 11 hypotheses are examined and discussed.

The final developed statistical models including the coefficient estimates for $\hat{UEI}_{Elec-adj}$ and $\hat{UEI}_{NG-adj}$ are depicted in Table 13.

**Table 13.** The final models' coefficient estimates.

| Regressor/Variable | Coefficient Estimates ($\hat{UEI}_{Elec-adj}$) | Coefficient Estimates ($\hat{UEI}_{NG-adj}$) | Both Energy Resources |
|---|---|---|---|
| Intercept | 5,1182.4 | 571.394 | |
| AWD | 2.6 | −0.00688 | *** |
| Vehicle Variety | 17 | 0.35536 | + |
| FM | −16 | −0.33441 | − |
| Number of Models | 69.7 | 0.26753 | + |
| Year | −22.3 | −0.25666 | − |
| Ln (APV) | −429.5 | −4.08135 | − |
| Ln (CAC) | −340.9 | −0.65277 | − |
| Plant Ownership (Japanese *) | −80 | −0.47642 | − |
| Vehicle Launch | 45.1 | . | *** |
| Segment B ** | 112.4 | 0.82821 | + |
| AC | 112.4 | NA | |
| Adjusted R-squared | 0.83 | 0.73 | |
| Cross-Validation—10 folds | 153 | 1.39 | |

* Japanese ownership is contrasted with American plants. ** Segment B is contrasted with segment A. *** The impact of all regressors on $\hat{UEI}_{NG-adj}$ and $\hat{UEI}_{Elec-adj}$ is similar other than these two variables.

Table 13 indicates that four regressors directly reduce UEIs and two regressors elevate them. Regressors that directly reduce UEIs include flexibility, APV, CAC, and production year. Observation year does not reduce UEIs per se; however, there are some latent factors such as technological improvement or work regime that might be confounding and occurring during the same period that may improve the estimated UEIs. On the contrary, vehicle variety and model types increase the UEIs. The study shows that while launching a new vehicle, the $UEI_{Elec\text{-}adj}$ increases; conversely, the $UEI_{NG\text{-}adj}$ remains unchanged during the launch. The paint line's ovens have the highest natural gas consumption in a plant. They are always running, except for the weekends or when the plants are down for the long holidays or maintenance activity. That could be considered as a probable reason that $UEI_{NG\text{-}adj}$ remains unchanged while launching a new product. In addition, another possible reason might be the low number of plants that had a new vehicle launch during the study period (n = 26).

The developed UEI equations are:

$$UEI_{Elec\_adj} = (51182.4 + 2.6AWD + 17\text{ Vehicle Variety} - 16\text{ Flexibility} + 69.7Model - 22.3Year - 429.5\ln(APV) - 340.9\ln(CAC) + 45.1Vehicle.launch - 80Japanese + 112.4Segment.B + 112.4AC) \quad (8)$$

$$UEI_{NG\_adj} = (571.394 - 0.00688AWD + 0.3553\text{Vehicle Variety} - 0.3344Flexibility + 0.2675Model - 0.2675Year - 4.08135\ln(APV) - 0.6527\ln(CAC) - 0.4764Japanese + 0.8282Segment.B) \quad (9)$$

The main findings from Table 13 and Equations (8) and (9) can be summarized for electricity and natural gas as follows:

Electricity:

- The regression model output indicates that 83% of the variation in $UEI_{Elec\text{-}adj}$ is explained by the group of 11 significant regressors.
- Cross-validation error is 153 kWh/vehicle.
- The $UEI_{Elec\text{-}adj}$ increases by 17 kWh or 2.26% (calculated based on using the mean of $UEI_{Elec\text{-}adj}$ = 752.96 kWh) for each unit increase in vehicle variety when all other regressors are fixed.
- When a new vehicle is launched, the $UEI_{Elec\text{-}adj}$ may increase by 45.1 kWh or 5.99%.
- The average $UEI_{Elec\text{-}adj}$ difference between vehicle segments A and B is 112.4 kWh or 14.93%.
- Japanese plants are more energy-efficient by consuming 80 kWh (10.62%) less electricity compared to American plants.
- The $UEI_{Elec\text{-}adj}$ is higher for plants that are cooled by 112.4 kWh or 14.93%.
- Natural gas:
- The regression model output indicates that 73% of the variation in $UEI_{NG\text{-}adj}$ is explained by the group of nine significant regressors; vehicle launch is not significant. However, the AWD variable impact is minimal and negligible.
- Cross-validation error is 1.39 MMBtu/vehicle (1 MMBtu ≈ 1055 MJ).
- The $UEI_{NGc\text{-}adj}$ increases by 0.356 MMBtu, 6.38% (calculated based on using the mean of $UEI_{NG\text{-}adj}$ = 5.58 MMBtu) for each unit increase in vehicle variety when all other regressors are fixed (1 MMBtu ≈ 1055 MJ).
- The average $UEI_{NG\text{-}adj}$ difference between vehicle segments A and B is 0.828 MMBtu or 14.84% (1 MMBtu ≈ 1055 MJ).
- Japanese plants are more energy-efficient by consuming 0.476 MMBtu (8.53%) less natural gas compared to American plants (1 MMBtu ≈ 1055 MJ).

Table 13 shows the impact of the developed/defined regressors on UEIs; hence, RQ2 is addressed. Moreover, the MRLM and hybrid models' coefficient estimates for both electricity and natural gas are compared in Tables A4 and A5, Appendix A, to understand what can be learned from the hybrid models. Tables A4 and A5 show:

- ○ Electricity: flexibility, vehicle launch, and plant ownership (Japanese) regressors are not significant at the 5% significance level in the MLRM; however, they are significant for the hybrid model, lasso with M-estimator (Huber). Moreover, the difference between the MLRM and hybrid methods' coefficient estimates for vehicle variety, plant ownership (Japanese), and vehicle launch variables is more than 20%.
- ○ sup>· Natural gas: AWD, vehicle segment, plant ownership, and number of models variables are not significant at the 5% significance level in the MLRM; however, they are significant for the hybrid model, lasso with MM-estimator. Moreover, the difference between the MLRM and hybrid methods' coefficient estimates for ln(CAC), plant ownership (Japanese), and Segment B variables is more than 20%.

The constructed hypotheses in Table 3 can be examined through the following test and considering Table 13 [64],

$$\begin{cases} H_0: \ \beta_i = 0 \text{ where i} = 1, 2, \ldots \text{P (P is the number of regressors, in this study 11)} \\ \qquad\qquad H_A: \ \beta_i \neq 0 \end{cases}$$

The null hypothesis in a regression model is represented by $t_0 = \hat{\beta}_j/\hat{se}(\hat{\beta}\,j) \sim t(n-p-1)$. The $H_0$ is rejected in favor of $H_A$ if $|t_0| > t(n-p-1, 1-\alpha/2)$. Table 14 summarizes the constructed hypotheses in Table 3 and illustrates their testing results.

**Table 14.** Hypothesis testing results.

| Research Hypotheses | ($\hat{UEI}_{Elec-adj}$) Hypotheses Testing Result (Y/N) | ($\hat{UEI}_{NG-adj}$) Hypotheses Testing Result (Y/N) |
|---|---|---|
| H1. Vehicle segment has a significant impact on the UEIs | Y—Segment A has the lower UEI | Y—Segment A has the lower UEI |
| H2. UEIs are correlated with CAC Utilization | Y | Y |
| H3. UEIs are correlated with the vehicle variety | Y | Y |
| H4. UEIs are correlated with the number of models | Y | Y |
| H5. UEIs are correlated with APV | Y | Y |
| H6. UEI variation for American and Japanese is significantly different | Y—Japanese has the lower UEI | Y—Japanese has the lower UEI |
| H7. UEIs are correlated with flexible manufacturing | Y | Y |
| H8. UEIs are correlated with AC | Y | NA |
| H9. UEIs are correlated with launching a new vehicle | Y | Y |
| H10. UEIs are correlated with AWD | Y | Y |
| H11. UEIs are correlated with year of production | Y | Y |

Table 14 shows that all 11 research hypotheses were accepted for both $\hat{UEI}_{Elec-adj}$ and $\hat{UEI}_{NG-adj}$ other than H9 for $\hat{UEI}_{NG-adj}$ (the potential reason for not accepting this hypothesis was discussed in Table 13 findings).

## 5. Conclusions and Implications

This section discusses the comparison of UEI and HPV determinants, policy implications, conclusions and study limitations, and future research.

### *5.1. UEI and HPV Determinant Comparison and RQ3 Examination*

In this subsection, the third research question

RQ3: What are the common determinants that can improve both productivity and energy efficiency KPIs at the same time?

is examined and discussed.

The paper outlines the underlying determinants of unit energy intensity, UEIs, and labor intensity, HPV. However, the study finds that these two KPIs are correlated, but the relationship is not one-to-one and the underlying determinants of these two KPIs can be different. Table 15 compares these two KPIs' (HPV and UEIs) coefficient estimates and addresses RQ3.

**Table 15.** Comparing HPV and UEI statistical models.

| | HPV *—(Coef. Sign) | $UEI_{Elec\text{-}adj}$ | $UEI_{NG\text{-}adj}$ |
|---|---|---|---|
| Intercept | + | 51,182.40 | 571.39 |
| AWD | + | 2.60 | −0.01 |
| Vehicle Variety | + | 17.00 | 0.36 |
| Number of Models | NA ** | 69.70 | 0.27 |
| Year | - | −22.30 | −0.26 |
| ln(APV) | - | −429.50 | −4.08 |
| ln(CAC) | + | −340.90 | −0.65 |
| Plant Ownership (Japanese) | NA ** | −80.00 | −0.48 |
| Vehicle Launch | + | 45.10 | . |
| Segment B | + | 112.40 | 0.83 |
| Segment C ** | + | NA ** | NA ** |
| FM | - | −16.00 | −0.33 |
| AC | NA ** | 112.4 | NA ** |

* Adapted from Abolhassani and Jaridi (2016) [53]. ** NA: Not Applicable/Available.

There are some factors that have similar effects on both HPV and UEIs. The AWD, FM, APV, and year of manufacturing enhance HPV and UEIs at the same time. On the contrary, vehicle variety, number of models, and launching a new vehicle penalize HPV and UEIs at the same time. Additionally, segment B is less productive and energy-efficient compared to segment A. The study shows Japanese manufacturers are more energy-efficient and productive compared to American OEMs. CAC is the only determinant that has an opposite sign when comparing HPV and UEIs.

The authors did not find any compelling reasons indicating that platform sharing and percentage of hourly employees can impact energy intensity, and consequently UEIs. Therefore, the authors decided not to include them in the UEI models.

The current study proposes a systematic and comprehensive approach to strengthen the automakers' energy efficiency. On one hand, it determines the correlation between labor intensity and energy intensity, HPV and UEIs, and studies whether or not car ownership, launching a new vehicle, and cooling a plant impact this correlation coefficient. On the other hand, it demonstrates several steps, including building base regression models involving various transformations, utilizing robust and regularized methods, and developing hybrid statistical models to estimate the UEIs. Our novel and hybrid statistical models outperform conventional multivariate regression models (the adjusted R-squared and mean squared cross-validation error for electricity improved from 0.77 to 0.83 and 188.5 to 153, respectively).

## 5.2. Practical and Policy Implications

Any plant operational manager could use the developed regression models, Equations (8) and (9), to obtain the preferred target values for UEIs. They can define the optimal values for each strategy by using Mixed Integer Programming (MIP) optimization and considering the plant's constraints. A metaheuristic optimization method or the branch and bound (BB) method can be utilized to obtain the optimal value for each strategy. The developed hybrid regression models represent the objective function and each regressor's limitation can be used as a constraint (lower or upper bound). For instance, for vehicle segment A, which was the most energy-efficient segment during the study period, with 568.24 kWh, the following MIP mathematical modeling can be used to attain a target $UEI_{Elec\text{-}adj}$ of 350 kWh (350 kWh is 38% energy efficiency enhancement compared to the most energy-efficient benchmark plant). Z is the distance from the specified target value for $UEI_{Elec\text{-}adj}$, 350 kWh in the example below.

$$Min(z) = (51182.4 + 2.6AWD + 17\ Vehicle\ Variety - 16\ Flexibility + 69.7Model - 22.3Year \\ -429.5\ln(APV) - 340.9\ln(CAC) + 45.1Vehicle.launch - 80Japanese + 112.4Segment.B \\ +112.4AC - 350)^2 \quad (10)$$

Subject to:

$234 \leq$ AWD, integer $\leq 283$;
$2 \leq$ Vehicle variety, integer $\leq 12$;
$13.8 \leq FM \leq 25.8$;
$1 \leq$ Number of car models, integer $\leq 5$;
$10.25 \leq \ln(APV) \leq 13.03$;
$2.56 \leq Ln(CAC) \leq 4.91$;
$1999 \leq$ Year, integer $\leq 2005$.

Z is the expected value of the Taguchi loss function and includes both the variance of the estimator and its deviation from the sample mean. Z can be determined as,

$$Z = (\overline{X} - T)^2 + S^2$$

where

Z = Taguchi loss function;
$\overline{X}$ = The sample mean;
T = The given target, in our research 350 kWh;
$S^2$ = The sample variance (since we are hitting the target in each iteration, the variance will be zero).

A nonlinear optimization model known as Mixed Integer Distributed Ant Colony Optimization, MIDACO, was utilized to overcome some common challenges such as convexity, flat spots, stochastic noise, and non-differentiability [69]. Table 16 depicts decision variables, decision variables' range, obtained optimum values, practical insight, and the mean squared error, Z, for segment A.

**Table 16.** Optimal regressors' values for $UEI_{Elec\text{-}adj}$ of 350 kWh—MIDACO Algorithm.

| Decision Variables | Decision Variables' Bounds ** | Optimum Attained Setting *** | Practical Insight |
|---|---|---|---|
| $UEI_{Elec\text{-}adj}$ | (345.55, 2380.17) | | |
| AWD * | (234, 283) | 240 | "AWD" is fixed at 240 |
| Vehicle variety | (2, 12) | 2 | Since "Vehicle variety" affects $UEI_{Elec\text{-}adj}$ negatively (its coefficient estimate has a positive sign), the attained optimum value is almost at the lower end of the spectrum, which means that the factory has to manufacture less vehicle variety |
| FM | (17.26, 24.26) | 21.58 | Since "FM" has a positive influence on $UEI_{Elec\text{-}adj}$ (FM coefficient estimate has a negative sign), the attained optimum value is almost inclined to the higher end of the spectrum, which means that the factory has to be more flexible |
| Number of Models | (1, 5) | 1 | Since "Model types" affects $UEI_{Elec\text{-}adj}$ negatively, the attained optimum value is at the lower end of the spectrum, which means that the factory has to manufacture fewer models |

**Table 16.** *Cont.*

| Decision Variables | Decision Variables' Bounds ** | Optimum Attained Setting *** | Practical Insight |
|---|---|---|---|
| LN(APV) | (10.25, 13.03) | 12.03 | Since "APV" has a positive influence on $UEI_{Elec\text{-}adj}$, the attained optimum value is almost at the higher end of the spectrum, which means that the factory has to manufacture at the greatest possible volume |
| LN(CAC) | (2.56, 4.91) | 4.01 | Since "CAC" has a positive influence on $UEI_{Elec\text{-}adj}$, the attained optimum value is almost at the higher end of the spectrum, which means that the factory has to manufacture at the greatest possible car assembly and capacity utilization (CAC) level |
| Year * | (1999, 2005) | 2005 | The "year" was fixed at 2005 |
| Vehicle launch | 1 | 1 | The "vehicle launch" was fixed at one, defining the optimal value when launching a new vehicle |
| Ownership | (American, Japanese) | Japanese | Since Japanese factories were more energy-efficient, Japanese ownership was selected (however, in reality, Japanese plants might possess a better management style, production system, technology, etc.) |
| Segment | A | A | Segment A was selected for the illustration in the optimization model |
| AC | 1 | 1 | It was assumed that the plant was cooled |
| | | Optimization model result | |
| KPI Measure | Target | Z | |
| $UEI_{Elec\text{-}adj}$ | 350 | $1.45 \times 10^{-18}$ | |

* The AWD and year were fixed at 240 and 2005, respectively. ** Each variable's bounds are shown in Table A6 in Appendix A (for segments A and B). *** The optimization solver was run for 100 iterations. Note: the reader should consider that some of the variables cannot be controlled by a plant manager.

The bio-inspired optimization method usually introduces stochastic noise, but the content of this research differs from the other optimization problems. The Z is calculated as $1.45 \times 10^{-18}$, which means that the deviations from target (350 kWh) are too small and negligible. Hence, the authors decided to accept the obtained regressors' values as optimal that were achieved through 100 iterations and did not perform a large number of iterations.

The authors used the final statistical model (Equation (8)) along with the regressors' range, Table A6, and examined a Monte Carlo simulation to explore if 350 kWh/vehicle is an appropriate target for $UEI_{Elec\text{-}adj}$. The histogram of 10,000 simulated $UEI_{Elec\text{-}adj}$ along with the 95% confidence interval are illustrated in Figure 4 (the 95% confidence interval ranges from 56.75 to 1745.26 kWh/vehicle). It was found that 10% of the simulated $UEI_{Elec\text{-}adj}$ were less than 350 kWh/vehicle; hence, 350 kWh was a reasonable target for $UEI_{Elec\text{-}adj}$.

The developed statistical model can be used to obtain the optimal settings for natural gas as well. Therefore, the OEMs can use the developed and implemented hybrid regression models along with mathematical modeling to improve their UEIs. While this approach can generally be applied to a manufacturing plant, it is essential to emphasize that the constraints and the regressors' range are specific to the companies being studied and could greatly impact the results for new plants.

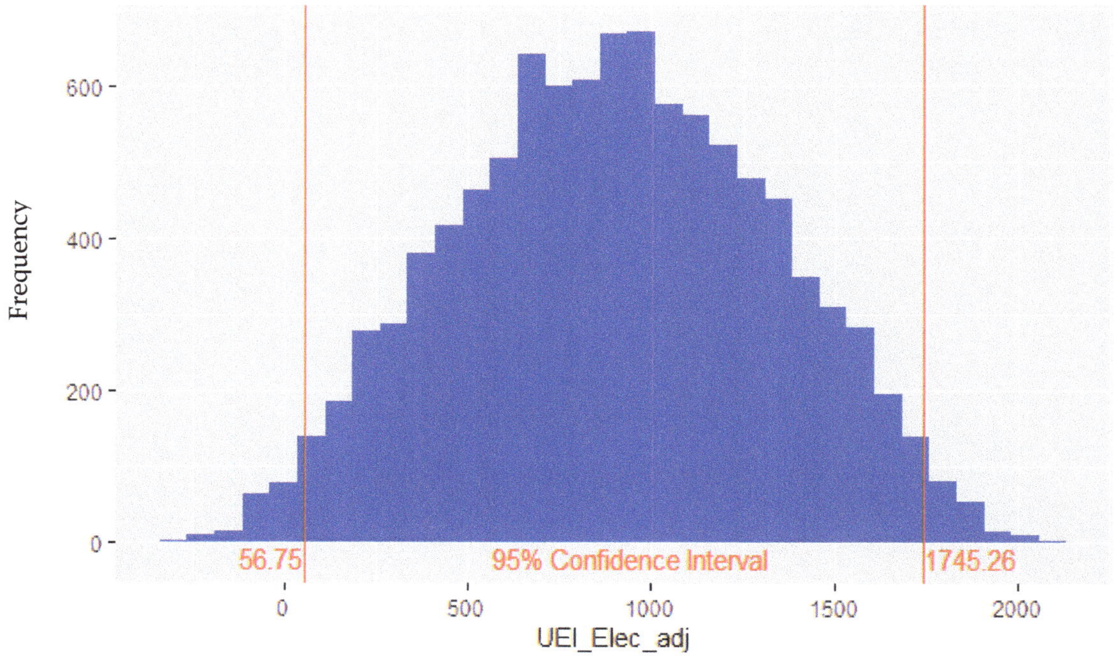


**Figure 4.** Simulated $UEI_{Elec\text{-}adj}$ histogram plot with 95% confidence interval.

### 5.3. Conclusions

The strategies to enhance energy intensity have become an important area of interest by many industries and researchers. In the literature, there are two common approaches for energy improvement: changing the industry frontier using the available best practices or implementing energy efficiency practices, such as replacing or upgrading the current equipment at a manufacturing plant [54]. The purpose of this research is different and complementary to the studies reported in the literature. There is little awareness of the group of regressors/variables that impact energy intensity, UEIs, at the theoretical and empirical levels in auto industry, increasing which is one of the key objectives of this study. Based on our analysis, the underlying determinants of energy efficiency and labor productivity are quite similar. This implies that strategies to improve energy may have spillover benefits to labor, and vice versa. The only underlying determinant that differs in sign is for capacity utilization. This may be due to the fact that labor is more flexible vis-à-vis utilization and energy has a larger fixed component.

In this study, we first investigated the association between energy intensity, UEIs, and labor intensity, HPV. The authors found that there was a statistically significant correlation between energy intensity and labor intensity for both electricity and natural gas (RQ1).

In the next step, the effect of 11 regressors on UEIs was examined, and the combination of M-estimator (Huber) or Multiple M-estimator and lasso regression was utilized to build robust statistical models to determine the impact of these regressors on UEIs. The regressors were chosen and developed based on a comprehensive literature review and practical purposes (RQ2).

It was found that launching a new vehicle and adding new car models were often seen as the primary reasons that would result in increased energy intensity. For instance, a one-unit increment in a new model adds a raise of 69.7 kWh, 9.26%, and 0.267 MMBtu, 4.781%, for UEIs (1 MMBtu $\approx$ 1055 MJ). Adoption of some complementary strategies in a manufacturing plant, such as increasing the CAC, FM, and production volume, was shown to reduce the UEIs.

Table 17 depicts an example for two different scenarios, and it indicates how vehicle variety could increase the $\hat{UEI}_{Elec-adj}$. In scenario 1, S1, assembling eight vehicle varieties would cost an extra 85 kWh per vehicle (=136 − 51 kWh) compared to scenario 2, S2, where a plant assembles three vehicle varieties.

**Table 17.** Vehicle variety strategy effect on $\hat{UEI}_{Elec-adj}$.

| | Vehicle Variety Coeff. per Table 13 | Number of Body Styles | Number of Chassis | Vehicle Variety | Effect on $\hat{UEI}_{Elec-adj}$ (kWh) |
|---|---|---|---|---|---|
| S1 | 17 | 3 | 5 | 8 | 136 |
| S2 | 17 | 1 | 2 | 3 | 51 |

The study shows flexibility is a crucial best practice that helps companies improve productivity and energy efficiency. Businesses should be flexible enough to respond effectively to various changes in today's competitive environment. Such flexibility can be achieved by creating a sustainable technology competency, improving employees' skills, and building an appropriate organizational culture that cultivates efficiency and effectiveness. One of the great examples of the flexible business model was when the COVID-19 pandemic hit the world. The impact on the companies which had a flexible business model was minimal and they were able to quickly adapt themselves during the Coronavirus shutdowns and reopening.

The authors also show that there is a strong alignment between energy efficiency and productivity determinants. They found that flexible manufacturing, production volume, and year of production can improve productivity and energy efficiency. On the contrary, vehicle variety, number of models, and launching a new vehicle penalized both productivity and energy efficiency (RQ3).

Japanese plants are more energy-efficient compared to American plants, for both electricity and natural gas energy resources. There might be some fundamental differences between Japanese and American car manufacturers that are worthwhile to consider. For example, American companies may produce less subcompact cars and produce more midsize SUVs, which consume more energy to produce. In addition, Japanese companies' subcompact cars may be smaller than American companies' subcompact cars, although they belong to the same vehicle segment. In addition, there may be some differences between Japanese and American OEMs' supply chain. For instance, the level of outsourcing might be different between Japanese and American plants. Furthermore, Japanese plants may intend to produce small parts inside and outsource large or semi-finished parts, which may not be the case for American plants.

It is worth noting that improvements in energy efficiency can often offset a car maker's profitability. For example, customers seem to prefer having more vehicle varieties or segment B vehicles (pickup, trucks, sport, luxury, and van) that may increase OEMs' production costs; however, it might be completely aligned with OEMs' profitability targets. Particularly in recent years, the variety of available vehicles to clients has significantly increased and customers prefer to have trucks or pickup cars (segment B) instead of small cars (segment A). Consequently, OEMs may intend to produce a greater number of car body types and chassis configurations specifically for segment B to increase their profitability at the expense of extra energy consumption. OEMs could use the proposed approach to define the optimal value of each strategy in order to simultaneously improve profitability and energy efficiency.

### 5.4. Study Limitations and Recommendations for Future Research

Similar to any research, there are limitations and data deficiencies in this study. For the plants that produced multiple vehicle segments, data were considered for the segment with highest volume. As this research was examined using the available data from 1999–2005, extrapolation to the current situation is not recommended, and the results must be viewed in the context of the available data. However, the principles behind the models and methodology will not change and can be replicated with a more recent dataset. The assumptions made for this research include:

- A total of 14% of $UEI_{Elec}$ was used for comfort cooling. It was assumed and found that the average energy share is linear in the range of CDD. The amount of 14% is only achieved when CDD is at the maximum, and $CDD_{adj}$ ranges from 0 to 1, which indicates that $UEI_{Elec\text{-}adj}$ ranges from 0.86 to 1. It means if plants A and B have the same UEI, but plant A is located in a hot climate (high CDD) and consequently cooled more, the UEI for plant A is lower than plant B, and plant A is more energy-efficient. On the other hand, if plant C has the lowest CDD and there is no electricity usage to cool the plant, all the electricity consumption was used to assemble vehicles.
- A total of 17% of $UEI_{NG}$ was used for comfort heating. It was found that the average energy share is almost linear in the range of HDD, and 17% is only achieved when HDD is at its maximum.
- The base temperature used to calculate CDD and HDD was 65 °F (≈18.3 °C).
- All plants were heated 24 h a day during the heating months.
- The plants were cooled 24 h a day during the cooling months, if cooled.
- The plants did not use on-site renewable energy sources or cogeneration and they purchased all their electrical energy.
- The authors were not able to exclude internal logistics, transportation, and supply energy consumption; hence, they were included in the study as plants' energy consumption.

Recommendations for possible future work include:

- Identifying new regressors that can be added to the UEI models. For example, the level of automation in a plant, plant's building size, and lean manufacturing may be included in the UEI models.
- The use of Group Lasso for the nominal variables (plant ownership and vehicle segmentation).
- In contrast to American plants, which failed to reduce UEIs over the study period, Japanese plants were able to reduce it. It might be worthwhile to study the reasons of increasing the UEIs for American plants during the study period. For instance, exploring the underlying determinants between Japanese and American management style might be of interest.
- Exploring the possible relationship between energy efficiency, productivity, sustainable smart manufacturing, sustainable industry 4.0, cyber-physical production network, and profitability [70,71].

**Author Contributions:** Conceptualization, B.G.; Methodology, G.B. and J.H.; Software, A.A.; Validation, A.A.; Formal analysis, A.A.; Resources, B.G.; Data curation, A.A.; Writing—original draft, A.A.; Writing—review & editing, G.B. and M.J.; Supervision, G.B., M.J. and J.H. All authors have read and agreed to the published version of the manuscript.

**Funding:** This research received no external funding.

**Data Availability Statement:** Due to the nature of this research, participants of this study did not agree for their data to be shared publicly, so supporting data are not available.

**Conflicts of Interest:** The authors declare no conflict of interest.

## Nomenclature

| | | | |
|---|---|---|---|
| UEI | Unit Energy Intensity | EPA | Environmental Protection Agency |
| HPV | Hours Per Vehicle | SME | Small and Medium Enterprise |
| MEP | Manufacturing Extension Partnership | EPI | Energy Performance Indicator |
| CDD | Cooling Degree Days | HDD | Heating Degree Days |
| GHG | Greenhouse Gas | DEA | Data Envelopment Analysis |
| IAC | Industrial Assessment Center | OEM | Original Equipment Manufacturers |
| APV | Annual Production Volume | CAC | Car Assembly and Capacity |
| AWD | Number of Annual Working days | FM | Flexible Manufacturing |
| MLRM | Multiple Linear Regression Model | AC | Air Cooling |

## Appendix A

### *Appendix A.1. Descriptive Statistics for $UEI_{Elec\text{-}adj}$ and $UEI_{NG\text{-}adj}$*

**Table A1.** Descriptive statistics for $UEI_{Elec\text{-}adj}$ and $UEI_{NG\text{-}adj}$.

| Yr. | Variable | Ownership | N | Mean | StDev | Min. | Q1 | Median | Q3 | Max. |
|---|---|---|---|---|---|---|---|---|---|---|
| 1999 | $UEI_{Elec\text{-}adj}$ | American | 25 | 650.9 | 315.8 | 370.5 | 471.5 | 631.4 | 859.4 | 1504.4 |
| | | Japanese | 2 | 538.5 | 29.3 | 500.8 | * | 538.5 | * | 542.3 |
| | $UEI_{NG\text{-}adj}$ | American | 25 | 5.185 | 2.657 | 1.935 | 3.453 | 5.013 | 7.104 | 10.999 |
| | | Japanese | 2 | 3.125 | 0.292 | 2.644 | * | 3.125 | * | 3.057 |
| 2000 | $UEI_{Elec\text{-}adj}$ | American | 23 | 675.9 | 383.4 | 381.1 | 529.2 | 630 | 881.1 | 1866.5 |
| | | Japanese | 2 | 525.28 | 3.48 | 522.26 | 522.26 | 525.28 | 528.29 | 528.29 |
| | $UEI_{NG\text{-}adj}$ | American | 23 | 5.185 | 3.178 | 1.795 | 3.548 | 5.036 | 7.218 | 16.264 |
| | | Japanese | 2 | 3.05 | 0.286 | 2.805 | 2.805 | 3.05 | 3.3 | 3.3 |
| 2001 | $UEI_{Elec\text{-}adj}$ | Japanese | 2 | 510.4 | 32.2 | 491.6 | * | 510.4 | * | 537.2 |
| | $UEI_{NG\text{-}adj}$ | Japanese | 2 | 2.929 | 0.248 | 2.454 | * | 2.929 | * | 2.805 |
| 2003 | $UEI_{Elec\text{-}adj}$ | American | 20 | 749.9 | 340.5 | 452.6 | 547.9 | 624.8 | 843.6 | 1885.3 |
| | | Japanese | 4 | 471.43 | 1.12 | 403.63 | 402.105 | 404.43 | 406.747 | 405.22 |
| | $UEI_{NG\text{-}adj}$ | American | 20 | 4.715 | 2.536 | 2.422 | 3.286 | 4.126 | 7.248 | 10.239 |
| | | Japanese | 4 | 2.78 | 0.308 | 2.16 | 2.187 | 2.378 | 2.573 | 2.596 |
| 2004 | $UEI_{Elec\text{-}adj}$ | American | 21 | 752 | 448.8 | 415.3 | 550.1 | 659.3 | 947.1 | 2380.2 |
| | | Japanese | 4 | 463.7 | 60.1 | 388.2 | 428.224 | 430.7 | 433.168 | 473.2 |
| | $UEI_{NG\text{-}adj}$ | American | 21 | 4.712 | 4.539 | 2.16 | 3.566 | 4.869 | 6.983 | 22.832 |
| | | Japanese | 4 | 2.752 | 0.59 | 2.074 | 2.292 | 2.492 | 2.696 | 2.909 |
| 2005 | $UEI_{Elec\text{-}adj}$ | American | 21 | 760.9 | 280.4 | 431.5 | 566 | 700.9 | 849.2 | 1622.6 |
| | | Japanese | 8 | 412.9 | 28.4 | 345.5 | 345.5 | 370.8 | 402.3 | 402.3 |
| | $UEI_{NG\text{-}adj}$ | American | 21 | 4.7 | 2.359 | 2.261 | 3.385 | 5.321 | 6.469 | 12.343 |
| | | Japanese | 8 | 2.605 | 0.242 | 2.029 | 2.029 | 2.406 | 2.48 | 2.48 |

### *Appendix A.2. Spearman's Correlation Test Statistics for the UEIs and Plant Size*

The UEIs do not follow the normal distribution; hence, the authors performed a non-parametric correlation test (Spearman's correlation) between the plant size and UEIs. The result of the Spearman's correlation test is illustrated in Table A2:

**Table A2.** Spearman's correlation test statistics for the UEIs and plant size.

| | Variables | Plant | $UEI_{Elec\text{-}adj}$ | $UEI_{Elec\text{-}adj}$ |
|---|---|---|---|---|
| All plants (n = 132) | Plant size | 1.0000 | | |
| | $UEI_{Elec\text{-}adj}$ | 0.3519<br>0.0000 | 1.0000 | |
| | $UEI_{NG\text{-}adj}$ | 0.2298<br>0.0014 | 0.7937<br>0.0000 | 1.0000 |

Table A2 indicates that there is a low degree of correlation between the plant size and UEIs. The authors believe that the size of a plant's buildings may have some impact on UEIs, but so far as the size was not available to the authors, they were not able to consider their impact on the UEIs. However, they tried to minimize the impact of comfort heating and cooling by adjusting the energy consumptions and using the $\widehat{UEI}_{Elec-adj}$ and $\widehat{UEI}_{NG-adj}$ (please see the research methodology section for more information).

*Appendix A.3. Fisher Transformation*

The correlations must be transformed for testing the difference between two independent coefficients. Each coefficient must be transformed with Fisher's r to z transformation [72]. The z transformation could be defined as

$$z = 0.5 \text{ Ln } \frac{1+r}{1-r}$$

(r is the correlation value).

To compare two correlations with different sample sizes $n_1$ and $n_2$, the Fisher's values can be calculated as $z_1$ and $z_2$, respectively. The population coefficients are equal under the null hypothesis and Z approximately follows the standard normal distribution and could be calculated as:

$$Z = \frac{z_1 - z_2}{\sqrt{\frac{1}{n_1-3} + \frac{1}{n_2-3}}}$$

Therefore, the correlation could be different between two samples at 5% level of significance, if the $|Z|$ is greater than or equal to 1.96. The Z value for different scenarios in Table 8 can be recalculated in Table A3.

**Table A3.** The correlations comparison for different scenarios.

| Different Group Comparison | UEI | Z |
|---|---|---|
| American vs. Japanese | $UEI_{Elec\text{-}adj}$ | 4.034 |
| American vs. Japanese | $UEI_{NG\text{-}adj}$ | 5.148 |
| Launch vs. no Launch | $UEI_{Elec\text{-}adj}$ | −1.276 |
| Launch vs. no Launch | $UEI_{NG\text{-}adj}$ | −2.521 |
| AC vs. no AC | $UEI_{Elec\text{-}adj}$ | −0.792 |
| AC vs. no AC | $UEI_{NG\text{-}adj}$ | −1.844 |
| GM vs. Ford | $UEI_{Elec\text{-}adj}$ | −2.382 |
| GM vs. Ford | $UEI_{NG\text{-}adj}$ | −1.470 |

Table AIII indicates (at 5% significance level):

- The UEI and HPV correlations are statistically different for the different plant characteristic, various ownership.

- The $UEI_{NG\text{-}adj}$ and HPV correlation is statistically different while launching a new vehicle.
- The $UEI_{Elec\text{-}adj}$ and HPV correlation is statistically different for Ford and GM.

*Appendix A.4. MRLM and Hybrid Method Coefficients' Estimate Comparison for $UEI_{Elec\text{-}adj}$ and $UEI_{NG\text{-}adj}$*

**Table A4.** MRLM and hybrid method coefficients' estimate comparison for $UEI_{Elec\text{-}adj}$.

| | MLRM | | Hybrid Method | |
|---|---|---|---|---|
| **Variable** | **Estimate** | **Pr(>\|t\|)** | **Estimate** | **% Change *** |
| Intercept | 47,839.52 | 0.00 | 51,182.40 | −6.99 |
| AWD | 3.07 | 0.00 | 2.60 | 15.19 |
| Variety.bodyandchassis | 11.13 | 0.02 | 17.00 | −52.77 |
| Model.types | 68.26 | 0.00 | 69.70 | −2.12 |
| Year | −20.53 | 0.00 | −22.30 | −8.61 |
| ln(APV) | −461.53 | 0.00 | −429.50 | 6.94 |
| ln(CAC) | −308.37 | 0.00 | −340.90 | −10.55 |
| Ownership.Japanese ** | −99.00 | 0.06 | −80.00 | 19.19 |
| Vehicle.Launch | 78.95 | 0.06 | 45.10 | 42.87 |
| Segment.B *** | 99.22 | 0.01 | 112.40 | −13.28 |
| AC | 121.80 | 0.00 | 112.40 | 7.72 |
| Flexibility | NA | NA | −16 | - |

* MRLM is the base. ** Japanese ownership is contrasted with American plants. *** Segment B is contrasted with segment A.

**Table A5.** MRLM and hybrid method coefficients' estimate for $UEI_{NG\text{-}adj}$.

| | MLRM | | Hybrid Method | |
|---|---|---|---|---|
| **Variable** | **Estimate** | **Pr(>\|t\|)** | **Estimate** | **% Change *** |
| (Intercept) | 578.5916 | $6.03 \times 10^{-5}$ | 571.394 | 1.24 |
| Variety.bodyandchassis | 0.36758 | $8.93 \times 10^{-7}$ | 0.35536 | 3.32 |
| Flexibility | −0.32597 | 0.037941 | −0.33441 | −2.59 |
| Year | −0.25658 | 0.000321 | −0.25666 | −0.03 |
| ln(APV) | −4.1223 | $5.24 \times 10^{-10}$ | −4.08135 | 0.99 |
| ln(CAC) | −2.33311 | 0.003436 | −0.65277 | 72.02 |
| Ownership.Japanese ** | −0.80878 | 0.086064 | −0.47642 | 41.09 |
| Segment.B *** | 0.56456 | 0.107781 | 0.82821 | 46.70 |
| Model.types | NA | NA | 0.26753 | – |
| AWD | NA | NA | −0.00688 | – |
| Vehicle.Launch | NA | NA | | - |

* MRLM is the base. ** Japanese ownership is contrasted with American plants. *** Segment B is contrasted with segment A.

*Appendix A.5. All Regressors'/Variables' Lower and Upper Bounds for Vehicle Segments A and B in Reference to the Available Data (1999–2005)*

**Table A6.** Ranges for $UEI_{Elec\text{-}adj}$, $UEI_{NG\text{-}adj}$, and predictors.

| Variable | Values Range | |
|---|---|---|
| **Segment** | **A** | **B** |
| $UEI_{Elec\text{-}adj}$ | (345.55, 2380.17) | (370.52, 1952.2) |
| $UEI_{NG\text{-}adj}$ | (1.77,23.85) | (2.49, 17.33) |
| AWD | (234, 283) | (183, 336) |
| Vehicle Variety | (2, 12) | (2, 20) |
| FM | (17.26, 24.26) | (16.77, 25.8) |
| Model types | (1, 5) | (1, 7) |
| APV | (28,356, 456,169) | (67,364, 426,499) |
| CAC | (13, 137) | (29, 150) |
| Year | (1999, 2005) | (1999, 2005) |
| AC | 0, 1 | 0, 1 |
| Launch of a new vehicle | 0, 1 | 0, 1 |
| Plant ownership | American, Japanese | American, Japanese |

## References


1. Sobral, M.C.; Jabbour, A.B.L.D.S.; Jabbour, C.J.C. Green Benefits from Adopting Lean Manufacturing: A Case Study from the Automotive Sector. *Environ. Qual. Manag.* **2013**, *22*, 65–72. [CrossRef]
2. Salleh, N.A.M.; Kasolang, S.; Jaffar, A. Green Lean Total Quality Information Management in Malaysian Automotive Companies. *Procedia Eng.* **2012**, *41*, 1708–1713. [CrossRef]
3. Bergmiller, G.G.; McCright, P.R. Are Lean and Green Programs Synergistic? In Proceedings of the Industrial Engineering Research Conference. In Proceedings of the Industrial Engineering Research Conference, Miami, CA, USA, 30 May–3 June 2009.
4. Boyd, G.A.; Pang, J.X. Estimating the linkage between energy efficiency and productivity. *Energy Policy* **2000**, *28*, 289–296. [CrossRef]
5. Admistration, E.I. Primary Energy Consumption by Source and Sector. 2014. Available online: http://www.eia.gov/totalenergy/data/monthly/pdf/flow/css_2014_energy.pdf (accessed on 21 December 2015).
6. Energy, U.S.D.O. Technology Road Map for Energy Reduction in Automotive. 2008. Available online: https://www1.eere.energy.gov/manufacturing/intensiveprocesses/pdfs/auto_industry_roadmap.pdf (accessed on 31 May 2017).
7. Hirsh, E. 2015 Auto Industry Trends. 2015. Available online: http://www.strategyand.pwc.com/perspectives/2015-auto-trends (accessed on 16 February 2016).
8. McKinsey. The Road to 2020 and beyond: What's Driving the Global Automotive Industry? 2013, pp. 1–24. Available online: https://www.mckinsey.com/~{}/media/mckinsey/dotcom/client_service/automotive%20and%20assembly/pdfs/mck_the_road_to_2020_and_beyond.ashx (accessed on 16 February 2022).
9. Wei, M.; Hong, S.H.; Alam, M. An IoT-based energy-management platform for industrial facilities. *Appl. Energy* **2016**, *164*, 607–619. [CrossRef]
10. Guradian, T. Why Automakers are the New Renewable Energy Investors. 2011. Available online: https://www.theguardian.com/sustainable-business/automakers-new-renewable-energy-investors-sustainable-cars (accessed on 16 February 2022).
11. Maroušek, J.; Gavurová, B. Recovering phosphorous from biogas fermentation residues indicates promising economic results. *Chemosphere* **2021**, *291*, 133008. [CrossRef] [PubMed]
12. Uludag, E.; Weisenstein, S. How to Reduce Hours Per Vehicle. *Automot. Des. Prod.* **2005**, *117*, 12–15.
13. Dorf, R.C. *The Technology Management Handbook*; CRC Press: Boca Raton, FL, USA, 1998.
14. Harbour. *The Harbour Report™ North America 2002*; Harbour: Detroit, MI, USA, 2002.
15. Kothari, C.R. *Research Methodology: Methods and Techniques*; New Age International: New Delhi, India, 2004.
16. Galitsky, C.; Worrell, E. *Energy Efficiency Improvement and Cost Saving Opportunities for the Vehicle Assembly Industry: An ENERGY STAR Guide for Energy and Plant Managers*; Lawrence Berkeley National Laboratorie: Berkeley, CA, USA, 2008. [CrossRef]
17. Kluczek, A. An energy-led sustainability assessment of production systems–An approach for improving energy efficiency performance. *Int. J. Prod. Econ.* **2019**, *216*, 190–203. [CrossRef]
18. Li, M.-J.; Tao, W.-Q. Review of methodologies and polices for evaluation of energy efficiency in high energy-consuming industry. *Appl. Energy* **2017**, *187*, 203–215. [CrossRef]

19. Feng, L.; Mears, L. Energy Consumption Modeling and Analyses in Automotive Manufacturing Plant. *J. Manuf. Sci. Eng.* **2016**, *138*, 101005. [CrossRef]
20. May, G.; Barletta, I.; Stahl, B.; Taisch, M. Energy management in production: A novel method to develop key performance indicators for improving energy efficiency. *Appl. Energy* **2015**, *149*, 46–61. [CrossRef]
21. May, G.; Stahl, B.; Taisch, M. Energy management in manufacturing: Toward eco-factories of the future–A focus group study. *Appl. Energy* **2016**, *164*, 628–638. [CrossRef]
22. Damert, M.; Baumgartner, R.J. Intra-Sectoral Differences in Climate Change Strategies: Evidence from the Global Automotive Industry. *Bus. Strategy Environ.* **2018**, *27*, 265–281. [CrossRef]
23. Boyd, G.A. Estimating the changes in the distribution of energy efficiency in the U.S. automobile assembly industry. *Energy Econ.* **2013**, *42*, 81–87. [CrossRef]
24. Boyd, G.A. Comparing the statistical distributions of energy efficiency in manufacturing: Meta-analysis of 24 Case studies to develop industry-specific energy performance indicators (EPI). *Energy Effic.* **2016**, *10*, 217–238. [CrossRef]
25. Oh, S.-C.; Hildreth, A.J. Estimating the Technical Improvement of Energy Efficiency in the Automotive Industry—Stochastic and Deterministic Frontier Benchmarking Approaches. *Energies* **2014**, *7*, 6196–6222. [CrossRef]
26. Jeon, H.W.; Taisch, M.; Prabhu, V.V. Modelling and analysis of energy footprint of manufacturing systems. *Int. J. Prod. Res.* **2014**, *53*, 7049–7059. [CrossRef]
27. Alsaleh, M.; Abdul-Rahim, A.S. Bioenergy Intensity and Its Determinants in European Continental Countries: Evidence Using GMM Estimation. *Resources* **2019**, *8*, 43. [CrossRef]
28. Alsaleh, M.; Abdul-Rahim, A.S. Bioenergy industry and the growth of the energy sector in the EU-28 region: Evidence from panel cointegration analysis. *J. Renew. Sustain. Energy* **2018**, *10*, 053103. [CrossRef]
29. Alsaleh, M.; Abdul-Rahim, A.S. Financial Development and Bioenergy Consumption in the EU28 Region: Evidence from Panel Auto-Regressive Distributed Lag Bound Approach. *Resources* **2019**, *8*, 44. [CrossRef]
30. Williams, J.; Alizadeh, R.; Allen, J.K.; Mistree, F. Using Network Partitioning to Design a Green Supply Chain. In Proceedings of the International Design Engineering Technical Conferences and Computers and Information in Engineering Conference, Online, 17–19 August 2020; American Society of Mechanical Engineers: New York, NY, USA, 2020; p. V11BT11A050.
31. Kliestik, T.; Zvarikova, K.; Lăzăroiu, G. Data-driven machine learning and neural network algorithms in the retailing environment: Consumer engagement, experience, and purchase behaviors. *Econ. Manag. Financ. Mark.* **2022**, *17*, 57–69.
32. Schulze, M.; Nehler, H.; Ottosson, M.; Thollander, P. Energy management in industry–a systematic review of previous findings and an integrative conceptual framework. *J. Clean. Prod.* **2016**, *112*, 3692–3708. [CrossRef]
33. May, G.; Stahl, B.; Taisch, M.; Kiritsis, D. Energy management in manufacturing: From literature review to a conceptual framework. *J. Clean. Prod.* **2017**, *167*, 1464–1489. [CrossRef]
34. Mardani, A.; Zavadskas, E.K.; Streimikiene, D.; Jusoh, A.; Khoshnoudi, M. A comprehensive review of data envelopment analysis (DEA) approach in energy efficiency. *Renew. Sustain. Energy Rev.* **2017**, *70*, 1298–1322. [CrossRef]
35. Prabhu, V.V.; Trentesaux, D.; Taisch, M. Energy-aware manufacturing operations. *Int. J. Prod. Res.* **2015**, *53*, 6994–7004. [CrossRef]
36. Bhattacharya, A.; Dey, P.; Ho, W. Green manufacturing supply chain design and operations decision support. *Int. J. Prod. Res.* **2014**, *53*, 6339–6343. [CrossRef]
37. King, A.A.; Lenox, M.J. Lean and Green? An Empirical Examination of the Relationship between Lean Production and Environmental Performance. *Prod. Oper. Manag.* **2009**, *10*, 244–256. [CrossRef]
38. Worrell, E.; Laitner, J.; Ruth, M.; Finman, H. Productivity benefits of industrial energy efficiency measures. *Energy* **2003**, *28*, 1081–1098. [CrossRef]
39. Inman, R.A.; Green, K.W. Lean and green combine to impact environmental and operational performance. *Int. J. Prod. Res.* **2018**, *56*, 4802–4818. [CrossRef]
40. Cherrafi, A.; ElFezazi, S.; Govindan, K.; Garza-Reyes, J.A.; Benhida, K.; Mokhlis, A. A framework for the integration of Green and Lean Six Sigma for superior sustainability performance. *Int. J. Prod. Res.* **2016**, *55*, 4481–4515. [CrossRef]
41. Diaz-Elsayed, N.; Jondral, A.; Greinacher, S.; Dornfeld, D.; Lanza, G. Assessment of lean and green strategies by simulation of manufacturing systems in discrete production environments. *CIRP Ann.* **2013**, *62*, 475–478. [CrossRef]
42. Chiarini, A. Sustainable manufacturing-greening processes using specific Lean Production tools: An empirical observation from European motorcycle component manufacturers. *J. Clean. Prod.* **2014**, *85*, 226–233. [CrossRef]
43. Herrmann, C.; Thiede, S.; Stehr, J.; Bergmann, L. An environmental perspective on Lean Production. In *Manufacturing Systems and Technologies for the New Frontier*; Springer: Berlin/Heidelberg, Germany, 2008; Volume 83–88.
44. Boyd, G.A.; Curtis, E.M. Evidence of an "Energy-Management Gap" in U.S. manufacturing: Spillovers from firm management practices to energy efficiency. *J. Environ. Econ. Manag.* **2014**, *68*, 463–479. [CrossRef]
45. Bloom, N.; Genakos, C.; Martin, R.; Sadun, R. Modern Management: Good for the Environment or Just Hot Air? *Econ. J.* **2010**, *120*, 551–572. [CrossRef]
46. Geng, R.; Mansouri, S.A.; Aktas, E. The relationship between green supply chain management and performance: A meta-analysis of empirical evidences in Asian emerging economies. *Int. J. Prod. Econ.* **2016**, *183*, 245–258. [CrossRef]
47. Wu, T.; Wu, Y.-C.J.; Chen, Y.J.; Goh, M. Aligning supply chain strategy with corporate environmental strategy: A contingency approach. *Int. J. Prod. Econ.* **2014**, *147*, 220–229. [CrossRef]

48. Zailani, S.; Jeyaraman, K.; Vengadasan, G.; Premkumar, R. Sustainable supply chain management (SSCM) in Malaysia: A survey. *Int. J. Prod. Econ.* **2012**, *140*, 330–340. [CrossRef]
49. Rath, B.N.; Akram, V.; Bal, D.P.; Mahalik, M.K. Do fossil fuel and renewable energy consumption affect total factor productivity growth? Evidence from cross-country data with policy insights. *Energy Policy* **2018**, *127*, 186–199. [CrossRef]
50. Rathilall, R.; Singh, S. Improving quality and productivity at an automotive component manufacturing organisation in Durban-South Africa. *Afr. J. Bus. Manag.* **2011**, *5*, 8854–8874.
51. Chen, F.W.; Tan, Y.; Chen, F.; Wu, Y.Q. Enhancing or suppressing: The effect of labor costs on energy intensity in emerging economies. *Energy* **2021**, *214*, 118964. [CrossRef]
52. Abolhassani, A.; Jaridi, M. Productivity enhancement in North American automotive industry: Strategies and techniques to reduce hours-per-vehicle. *Int. J. Product. Perform. Manag.* **2016**, *65*, 1112–1136. [CrossRef]
53. Snee, R.D. Validation of regression models: Methods and examples. *Technometrics* **1977**, *19*, 415–428. [CrossRef]
54. Chok, N.S. Pearson's versus Spearman's and Kendall's correlation coefficients for continuous data. In *Public Health in Partial Fulfillment*; University of Pittsburgh: Pittsburgh, PA, USA, 2010.
55. Royston, J.P. An Extension of Shapiro and Wilk's W Test for Normality to Large Samples. *J. R. Stat. Soc. Ser. C (Appl. Stat.)* **1982**, *31*, 115–124. [CrossRef]
56. Kohavi, R. A study of cross-validation and bootstrap for accuracy estimation and model selection. *Int. Jt. Conf. Artif. Intell.* **1995**, *14*, 1137–1145.
57. Dettling, M. *Applied Time Series Analysis*; Elsevier: Amsterdam, The Netherlands, 2013.
58. Lu, G.; Ding, X.; Peng, D.X.; Chuang, H.H. Addressing endogeneity in operations management research: Recent developments, common problems, and directions for future research. *J. Oper. Manag.* **2018**, *64*, 53–64. [CrossRef]
59. Fox, J.; Weisberg, S. *An R Companion to Applied Regression*; Sage: London, UK, 2010.
60. Cody, R. *Introduction to SAS University Edition*; SAS Institute: Care, NC, USA, 2015.
61. Alma, Ö.G. Comparison of robust regression methods in linear regression. *Int. J. Contemp. Math. Sci.* **2011**, *6*, 409–421.
62. Neter, J.; Kutner, M.H.; Nachtsheim, C.J.; Wasserman, W. *Applied Linear Statistical Models*; McGraw-Hill International: Chicago, IL, USA, 2005; Volume 4.
63. Susanti, Y.; Pratiwi, H.; Sulistijowati, S.; Liana, T. M Estimation, S Estimation, And Mm Estimation in Robust Regression. *Int. J. Pure Appl. Math.* **2014**, *91*, 349–360. [CrossRef]
64. Yohai, V.J. High Breakdown-Point and High Efficiency Robust Estimates for Regression. *Ann. Stat.* **1987**, *15*, 642–656. [CrossRef]
65. Koller, M.; Stahel, W.A. Sharpening Wald-type inference in robust regression for small samples. *Comput. Stat. Data Anal.* **2011**, *55*, 2504–2515. [CrossRef]
66. Hoerl, A.E.; Kennard, R.W. Ridge regression: Biased estimation for nonorthogonal problems. *Technometrics* **1970**, *12*, 55–67. [CrossRef]
67. Hastie, T.; Tibshirani, R.; Friedman, J.H.; Friedman, J.H. *The Elements of Statistical Learning: Data Mining, Inference, and Prediction*; Springer: Berlin/Heidelberg, Germany, 2011.
68. James, G.; Witten, D.; Hastie, T.; Tibshirani, R. *An Introduction to Statistical Learning*; Springer: New York, NY, USA, 2013; Volume 112.
69. Schlueter, M. MIDACO software performance on interplanetary trajectory benchmarks. *Adv. Space Res.* **2014**, *54*, 744–754. [CrossRef]
70. Andronie, M.; Lăzăroiu, G.; Iatagan, M.; Hurloiu, I.; Dijmărescu, I. Sustainable Cyber-Physical Production Systems in Big Data-Driven Smart Urban Economy: A Systematic Literature Review. *Sustainability* **2021**, *13*, 751. [CrossRef]
71. Durana, P.; Perkins, N.; Valaskova, K. Artificial intelligence data-driven internet of things systems, real-time advanced analytics, and cyber-physical production networks in sustainable smart manufacturing. *Econ. Manag. Financ. Mark* **2021**, *16*, 20–30.
72. Kenny, D.A. *Statistics for the Social and Behavioral Sciences*; Little, Brown: Boston, MA, USA, 1986.

energies

MDPI

*Article*

# Experimental Investigation of Impacts of Initial Pressure Levels on Compression Efficiency and Dissolution in Liquid Piston Gas Compression

**Barah Ahn and Paul I. Ro** *

Department of Mechanical Engineering, Baylor University, Waco, TX 76798, USA
* Correspondence: paul_ro@baylor.edu

**Abstract:** Understanding how the pressure level affects the efficiency of liquid piston gas compression is essential for a greater applicability of the technology in compressed air energy storage. To explore the impacts, compression starting at three different initial pressure levels (1, 2, 3 bar) with a pressure ratio of 2 is performed, and how isothermal compression efficiencies are affected depending on the initial pressures is analyzed. Under the experimental conditions, higher initial pressure leads to lower isothermal efficiency. Air dissolution during the compression is also investigated because the chamber is a pressure-varying and a liquid-containing environment, where the gas solubility changes during the process. Evaluating the dissolution is critical as it affects the energy output when the compressed air is expanded to regenerate the energy. The changes in the air mass and the retrievable volume of the air after expansion are quantified based on Henry's law. For a compression at higher pressure, because the air solubility is proportional to pressure, a greater reduction in the air mass and volume percentages is expected. This trend of the mass decreasing with the pressure level leads to less energy output than the originally intended output when the stored energy is retrieved in a discharging process.

**Keywords:** compressed air energy storage; liquid piston; multistage gas compression; compression efficiency

## 1. Introduction

Climate change is one of the most urgent issues that require much attention. For some developing countries that heavily depend on agriculture, it can bring about a heavier burden on their population [1]. Promoting the use of renewable energy sources can reduce fossil fuel use and curtail greenhouse gas emissions [2], which will lead to curbing climate change. However, their intermittent nature is a challenging issue when it comes to exploiting renewable energy, and building an energy storage system can help overcome the issue [3,4]. Storing energy when the demand for energy is low and distributing the stored energy to the population in on-peak hours enables a consistent supply of energy.

Compressed air energy storage (CAES) is one of the few bulk-scale energy storage options that are already commercialized. Due to the economic and environmentally friendly features of CAES, it is perceived as a promising choice for energy storage [5,6]. The fundamental mechanism of CAES is to store energy in the form of compressed air. During peak hours, the stored compressed air is expanded to run a turbine to restore the energy [7,8]. As a gas is compressed, heat is generated, whereas heat is lost when a gas expands. Depending on how CAES treats the heat generated during compression, it can be classified into three different types: diabatic CAES (D-CAES), adiabatic CAES (A-CAES), and isothermal CAES (I-CAES) [9,10]. In D-CAES, compression heat is dissipated before the compressed air is transferred to a storage vessel. In the discharging phase to regenerate the stored energy, an external energy source such as natural gas needs to be brought in to increase the temperature of the gas because gas expansion accompanies a significant temperature drop [9,11].

In D-CAES, the heat dissipation leads to the loss of energy. The idea of A-CAES stems from saving the heat energy that is wasted in conventional D-CAES. In A-CAES, a thermal energy storage device stores the compression heat energy generated during the compression step. In the discharging process, the stored heat can be employed to increase the gas temperature instead of using an external energy source. Therefore, it reduces the energy waste and eliminates the need for extra heat supply [9]. I-CAES is considered to be the most efficient among the three types [9,10]. By maximizing heat transfer, the temperature change is minimized during compression/expansion [12,13]. Because no energy is compromised by the heat generation, less work input is needed to compress air [12]. In addition, it does not require an extra device such as a thermal energy storage device of A-CAES, and no extra energy input is needed for the expansion stage. Owing to its promising characteristics, a variety of I-CAES models have been studied. Researchers in [14] proposed a system equipped with a near-isothermal compression system for wind energy, which serves as one of the major electricity generation sources in the US [15] and is rapidly growing globally [16]. According to their quantified evaluation, the suggested model resulted in the size and weight reduction, which leads to cost saving [14]. Researchers in [17] focused on the expansion process where the energy discharging takes place. Compared to adiabatic expansion, the introduction of their near-isothermal expander resulted in a 15.7% increase in specific work.

The liquid piston gas compression concept was recently re-introduced by Van de Ven and Li [12]. The fundamental mechanism of the compressor is that a liquid column functions as the piston of a conventional reciprocating compressor. It has been drawing attention as an innovative compression technique for energy storage applications as a variety of heat transfer improvement techniques can be utilized in the compressor, owing to the conforming nature of a liquid that can take a complicated shape. In other words, it has a great potential to achieve a near-isothermal process. A number of experimental and simulation studies have been performed to investigate the heat transfer improvement techniques in a liquid piston compressor. Placing an insert is a direct way to provide extra surface area for heat transfer. The effects of porous media and interrupted-plate-type inserts were examined in [18]. For the same level of power density in their test, a 18% efficiency increase was achieved with the inserts. According to [19], at the much higher target pressure of 210 bar, compression efficiency improvement was observed with the use of porous media. Metal wire mesh inserts were introduced as an insert for the liquid piston in [20]. An increase of 6–8% in compression efficiency with the metal wire inserts was attained over the baseline liquid piston compression. By testing two different materials, aluminum and copper, the impacts of the thermal property of an insert were also studied, and it turned out that the material does not make a notable efficiency change. Metal plates were also used as an insert in [21]. By varying the compression speed and the dimensions of inserts, a compression efficiency improvement of up to 8% was observed under their experimental conditions. Liquid-based heat transfer techniques have also been studied. Researchers in [13] investigated factors that affect droplet heat transfer by varying the mass of injected water and droplet size. To some degree, reducing the size of droplets is beneficial, and enlarging the mass of droplet injection is advantageous for boosting heat transfer. Researchers in [22] experimentally evaluated the spray injection technique by varying the angle of the nozzle and spray injection pressure. Higher injection pressure leads to higher flowrate and introduces a greater amount of water into the compression chamber. Thus, less temperature increase during compression was observed. However, as higher injection pressure requires higher work input, a higher spray pressure does not always guarantee a better overall efficiency despite the smaller temperature increment. Researchers in [23] examined the effectiveness of utilizing two proven techniques of metal inserts and spray injection at a time. They performed a more detailed analysis of the impacts of spray injection pressure on droplet heat transfer at different injection pressures. They proved that the spray pressure is critical in determining droplet heat transfer, including the size and velocity of a droplet as well as spray flowrate, all of which have an impact on droplet heat transfer and are dependent on the pressure. Aqueous foam is another liquid-based heat transfer technique, and an experimental investigation was carried out in [24]. Aqueous foam is generated inside the compression chamber so that extra heat transfer surface area is provided, and it increases the isothermal efficiency to 92% from 86% of the compression without the foam. Compared to the way spray injection is employed in [22,23], where the nozzle is continuously in operation throughout the compression, foam generation is carried out once before the compression begins. Therefore, it is expected to save work input compared to the other liquid-based method. A similar gas compression technique that utilizes a liquid and a porous medium was introduced and experimentally assessed in [25,26]. They examined the issue of gas dissolution during compression, which needs to be addressed when a liquid is used for compression. Using Henry's law, the amount of dissolved air was estimated based upon a proportional relationship between the solubility and the pressure. A scaled-up liquid piston has also been tested in [27]. A chamber with a volume of 24.71 $m^3$ was designed, and exergy-based efficiency evaluation was performed under various experimental conditions. On top of bringing external heat transfer enhancement, a variety of approaches to optimizing the liquid piston itself have also been made. In [28], chambers designed with diverse design parameters were examined. Because the shape of the chamber affects the flow behavior of the internal air, the change in the chamber shape can result in heat transfer efficiency variations. In [29], compression speed was varied and different liquid piston chambers were experimentally investigated. The faster compression leads to lowered isothermal performance, and the thermal conductivity of the chamber material has a relatively small impact on heat transfer performance. A chamber that consists of a number of tube arrays was also tested [30]. Increasing the number of tubes leads to a positive effect on compression efficiency.

Many experimental studies on liquid piston compression focus on compression starting from atmospheric pressure or compression with a single case of pressure variation. In the current work, liquid piston gas compression at different initial pressure levels is tested, and how the different pressure affects compression efficiency for the same pressure ratio is examined. As multi-stage compression can lead to a better efficiency than single-stage compression [31], and the CAES plants in the actual operation employ multi-stage compressors [9,32,33], this work can aid in evaluating the real applicability of a liquid piston at an elevated pressure in a multi-stage system. Furthermore, this can pave the way for the future application of pressure-sensitive techniques such as spray injection and aqueous foam by broadening the fundamental understanding of liquid piston efficacy at various pressures. In addition, the dissolution issue of a liquid piston is also investigated by estimating how much air dissolution takes place for each process. It is also identified whether the difference in the dissolution contributes to the compression efficiency. There have been separate studies that examined the impact of the initial pressure [27,30] or identified the dissolution issue [25,26,34,35] in a liquid piston or similar liquid-based compression systems. To the author's best knowledge, this is the first work that relates the impacts of different pressure levels on liquid piston compression efficiency and air dissolution. Because this work examines relatively moderate pressure ranges, which were not covered in [27,30] and where air solubility data can be easily estimated with the data available in [36], those two issues that are closely associated can be studied simultaneously. Another novelty of the current work is to investigate how the dissolution affects the overall efficiency of the energy storage system. In addition, by taking into consideration the discharging process of CAES, the study allows for a practical evaluation of a liquid piston system for energy storage applications.

This paper consists of five sections. The current section introduces the background of this research. The following section illustrates the analytical modeling to describe the concept and the methodology with the experimental apparatus description. The next section presents the results of the experiments. Then, analysis of the results is presented in the following section. Lastly, the paper ends with concluding remarks.

## 2. Materials and Methods

### 2.1. Analytical Modeling

The mathematical model of the liquid piston system used in this work can be represented in the same way as [20,22–24,29]. Using the first law of thermodynamics, the energy equation for liquid piston gas compression can be written as Equation (1).

$$\dot{U} = \dot{Q} - \dot{W}_{comp} \tag{1}$$

where $\dot{U}$, $\dot{Q}$, and $\dot{W}_{comp}$ indicate the rate of internal energy, heat transfer, and compression work, respectively. Each term has an alternative form as follows:

$$\dot{U} = m_{gas} C_v \frac{dT_{gas}}{dt} \tag{2}$$

$$\dot{Q} = U_h A_s (T_\infty - T_{gas}) \tag{3}$$

$$\dot{W}_{comp} = P_{gas} \frac{dV_{gas}}{dt} \tag{4}$$

By replacing the terms with the alternative forms, Equation (1) can be rewritten as Equation (5):

$$m_{gas} C_v \frac{dT_{gas}}{dt} = U_h A_s (T_\infty - T_{gas}) - P_{gas} \frac{dV_{gas}}{dt} \tag{5}$$

From Equation (5), $-\dot{W}_{comp}$ is positive due to the volume decrease, and $\dot{Q}$ is negative because the gas temperature increases and becomes higher than the environmental temperature during compression. Hence, the right-hand side is equivalent to the difference in the absolute values of the work rate term and the heat transfer term. Equation (5) can be rearranged in terms of the gas temperature change.

$$\frac{dT_{gas}}{dt} = \frac{\left|\dot{W}_{comp}\right| - \left|\dot{Q}\right|}{m_{gas} C_v} \tag{6}$$

Generally, in gas compression, $\left|\dot{W}_{comp}\right|$ is greater than $\left|\dot{Q}\right|$. By increasing heat transfer, the numerator of the right-hand side of Equation (6) becomes closer to zero. This results in a smaller temperature change rate on the left-hand side, bringing it closer to an isothermal compression with a reduced work input.

Figure 1 was motivated by [31] and displays the pressure–volume trajectories of various gas compression processes starting from the initial point of $(V_0, P_0)$ to the compression end point of $\left(V_{f,iso}, P_f\right)$. Figure 1a compares typical trajectories of isothermal, adiabatic, and polytropic compression processes. Equation (7) gives the polytropic process expression [37]:

$$P_1 V_1^n = P_2 V_2^n \tag{7}$$

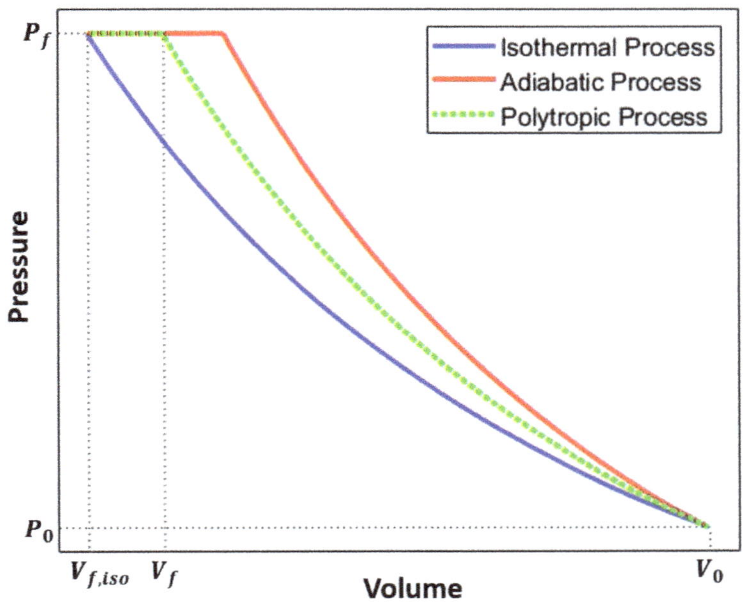


(a)

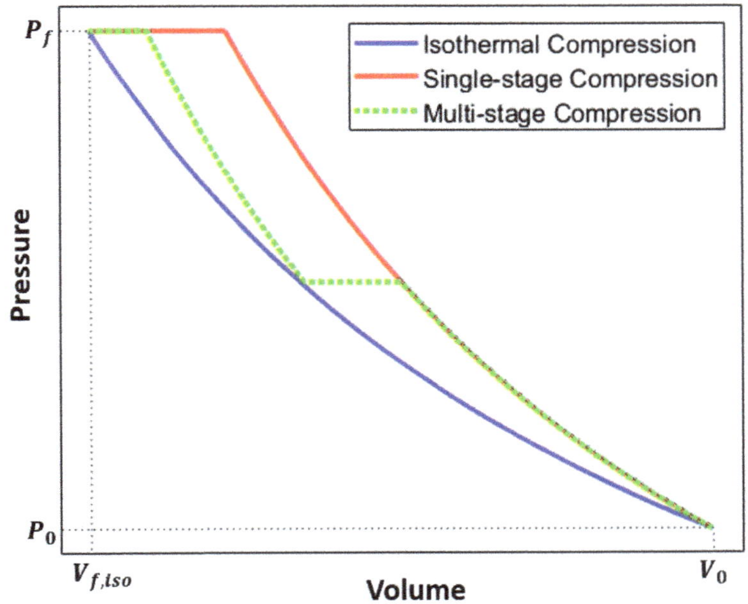


(b)

**Figure 1.** Compression Trajectories Comparison of (**a**) Isothermal, Adiabatic, Polytropic Processes; (**b**) Isothermal, Single-stage, Multi-stage Compression.

For the isothermal process, the polytropic index $n$ of Equation (7) is 1, while it is 1.4 for the adiabatic process [10]. If the index has a closer value to 1, the trajectory labeled as the Polytropic Process moves closer to the isothermal process trajectory. For a non-isothermal process, it attains the desired pressure $P_f$ at $V_f$ before the gas volume reaches the desired volume of $V_{f,iso}$. Thus, extra cooling work is needed to move from $V_f$ to $V_{f,iso}$ [22,29,38].

The amount of work input can be determined by calculating the area under its trajectory. Thus, the isothermal process takes less work input to complete the compression compared to non-isothermal processes, and less energy is required to perform compression if the polytropic index is smaller and closer to the isothermal process, as shown in Figure 1a. The isothermal efficiency of a certain process can be obtained with Equation (8) where $Pr$ is the pressure ratio, $P_f/P_0$ [29,38].

$$\eta_{iso} = \frac{W_{isothermal}}{W_{actual}} = \frac{\overbrace{mRT_0\left(lnPr - 1 + \frac{1}{Pr}\right)}^{W_{isothermal}}}{\underbrace{-\int_{V_0}^{V_f}(P(V) - P_0)dV}_{W_{comp}} + \underbrace{P_0(Pr-1)\left(V_f - \frac{V_0}{Pr}\right)}_{W_{cooling}}} * 100\ [\%] \tag{8}$$

Figure 1b straightforwardly shows why multi-stage compression can be more efficient than single-stage compression. The two trajectories have the same polytropic index. However, by pausing in the middle of the process and allowing for cooling down, the multi-stage compression consumes less work input than the single-stage compression. Although an actual cooling process may not follow an ideal flat profile as it is shown in Figure 1b, it gives an idea on how multi-stage compression can be advantageous.

Gas dissolution needs to be identified because the liquid piston chamber is a pressure-varying environment that contains water. The dissolution affects multiple aspects of evaluating the liquid piston technique. It complicates the overall gas temperature estimation during the compression. Many experimental studies of the liquid piston employed a thermocouple to directly measure the temperature of the gas during the process [20,22–24,29,39]. However, because the measured temperature is local to the tip of the thermocouple, the overall temperature of the air inside the chamber cannot be represented with the measured value. The ideal gas equation can be used to calculate the overall temperature with the simple modification as Equation (9).

$$T = \frac{PV}{mR} \tag{9}$$

Previous simulation-based liquid piston studies used the ideal gas equation to estimate the temperature of the next step by combining the polytropic process relationship of Equation (7) [12,13].

$$T_i = \frac{T_{i-1}V_{i-1}^{n-1}}{V_i^{n-1}} \tag{10}$$

However, if air dissolves and the mass of the air changes during the process, the temperature obtained based on the ideal gas equation leads to an error. This results in a circular problem in this study when it comes to estimating the gas temperature and the dissolved mass with the ideal gas equation. The mass of the air is needed to calculate the air temperature, while the mass calculation requires the air temperature. Hence, the quantification of the gas mass changes during compression is highly implausible with the ideal gas equation. However, the quantification of the dissolved gas is critical because it affects the compression work input and the energy that can be restored. Employing the proportional relationship of the gas solubility and pressure is a proper approach to estimating the dissolved gas. Henry's law illustrates the proportional relationship of the gas dissolution [36].

$$S_{gas} = \frac{P_{gas}}{k_H} \tag{11}$$

Even with Henry's law, it is still highly difficult to perform a real-time quantification of the dissolved gas because the basic form of Henry's law equation has a single variable for pressure while temperature affects the solubility values as well. Therefore, it is difficult to make a proper adjustment to the equation for every data point. Hence, simplified estimations with solubility data available on [36] are made. The total reduced work from the dissolution can be calculated using the following equation, which calculates the area under the linear profile of the solubility and pressure.

$$W_{dis} = \left(P_f - P_0\right) * \left(S_f - S_0\right) * V_w/2 \tag{12}$$

The dissolution is advantageous for the isothermal compression efficiency calculation with Equation (8), whereas the mass loss due to the dissolution during compression cannot be restored back outside the chamber and compromises the amount of energy that can be restored. To evaluate how much work is compromised when the energy is discharged, the mass loss from the dissolution needs to be calculated. Assuming the air follows the ideal gas law, the air mass loss from the dissolution is quantified by employing the solubility data and the ideal gas equation.

### 2.2. Experimental Apparatus

A liquid piston gas compressor setup was built for the current study. Figure 2 presents a schematic diagram of the system and the actual picture of the compression chamber.

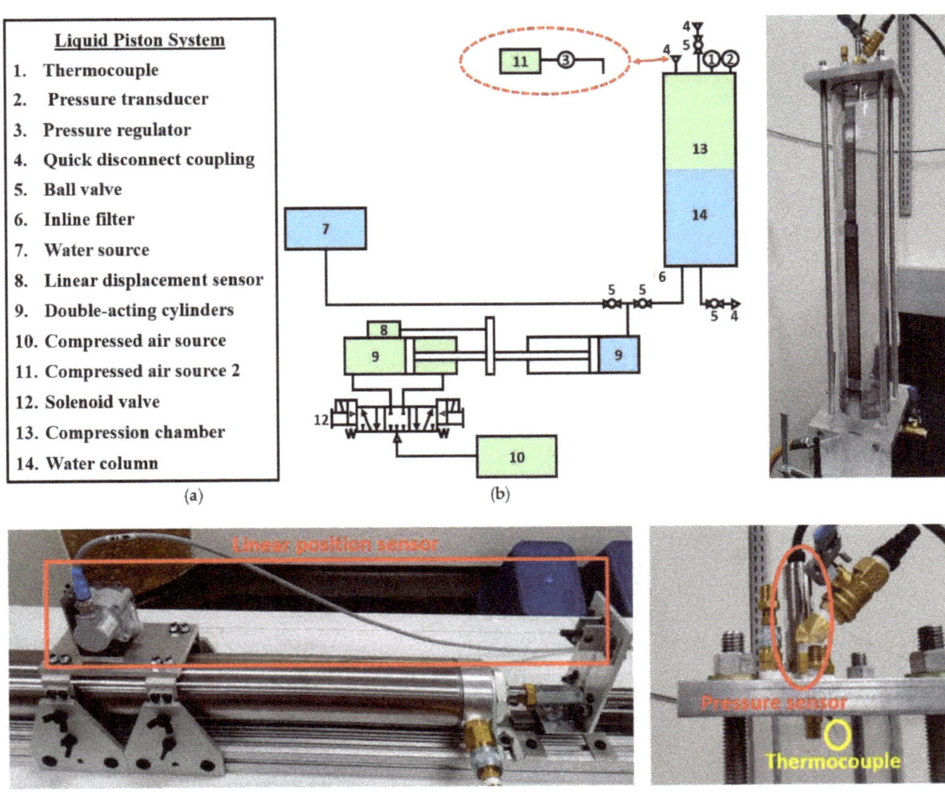


**Figure 2.** Liquid Piston Setup: (**a**) Schematic Diagram; (**b**) Actual Compression Chamber; (**c**) Sensor Configuration.

The schematic diagram of the overall system configuration is depicted in Figure 2a. A 2-foot-long transparent polycarbonate cylinder sandwiched by two aluminum plates serves as a compression chamber (#14 in the diagram). The inner diameter is 3.5 inches and the outer diameter is 4 inches. The initial water level is set slightly above the bottom of the chamber so that the height of the gas-filled region is about 22.5 inches. Two pneumatic cylinders (#9) function as a water pump to convey water into the chamber. As one of the cylinders (#9 left) connected to a power source (#10) extends, the water residing in the other cylinder (#9 right) directly coupled with the chamber is pushed into the chamber. As the water enters the chamber and the water level increases, the air inside the chamber is compressed. To measure the volume of the gas, a position sensor (#8) is installed on the water pumping system. A thermocouple (#1), pressure transducer (#2), a quick-disconnect valve coupler (#4), and an on/off valve (#5) are placed on the cap of the chamber. The quick-disconnect coupler is located 1 inch away from the axial center of the top plate and is used to set or initialize the pressure of the gas. For experiments that require higher initial pressures, air is injected through the coupler and the gas pressure rises with the red-circled parts connected to another compressed air injection source (#11). When the pressure level needs to be initialized to the atmospheric pressure, a quick-disconnect plug is used to open the coupler, which works as an outlet to the atmosphere. On the ceiling of the chamber at the axial center, a spray nozzle is positioned for the integration of the droplet injection heat transfer technique, which is not used in this current study. Figure 2c shows the configuration of the sensors. The thermocouple and the pressure transducer are located 1 inch away from the axial center of the cap of the chamber and directly measure the temperature and pressure of air. The linear position sensor that measures the displacement of the rod end of the cylinders enables the measurement of air volume. By measuring the position, the amount of water entering the chamber can be calculated. The water volume entering the chamber means the change in the volume of air inside the chamber. As it is an indirect way to measure the volume and it may include some machining error during the installation, additional tests were carried out to compare the water volume changes calculated from the sensor and the changes in the liquid piston level. For the water level, an error of around or higher than 1% was observed in the early stage of the process. Near the end of the process, it was mostly less than 1%. The manufacturer, model name, and accuracy information of the sensors used in this study are summarized in Table 1.

**Table 1.** Sensors Used in Experimental Setup.

| Sensor | Model | Accuracy |
|---|---|---|
| Pressure transducer | Omega Engineering PX409-250A10V | ±0.08% BSL accuracy |
| Thermocouple | Omega Engineering 5TC-TT-K-40-72 | 2.2 °C |
| Position sensor | TE SPD-50-3 | 0.25% |

To examine the impacts of the initial pressure levels on isothermal efficiency, air compression processes were performed for three different initial pressure levels of atmospheric pressure, which was about 1 bar, 2 bars, and 3 bars. The target pressure ratio was 2 for all the three pressure cases. Compression was carried out for a certain time until a pressure ratio higher than 2 was reached to be safe. Then, the profiles from the starting point to the pressure ratio of 2 were compared. By adjusting the water pump power for each case, the stroke times taken to reach the target ratio were closely matched. For the process at the atmospheric pressure, once the compression was finished, the outlet was manually opened to initialize the pressure level to the atmospheric pressure. Then, the liquid piston was retracted, resulting in water taken out of the chamber. For the compression starting from 2 and 3 bars, all the procedures were automated using an Arduino controller. The initial pressure was set-up by injecting air through the valve on the top plate (#4), and a pressure regulator adjusted the pressure to the desired level before performing compression. Due to the manual operation, the initial pressure levels were not exactly 2 and 3 bars. Once the compression was finished, the expansion was performed over a few steps to prevent the sudden temperature drop in a short time, which may not be desirable for the sensor operations. Throughout these processes with the elevated initial pressure, the outlet did not open to the outside and the system remained closed for the constant starting pressure levels. This process was repeated six times and the last five of them were averaged for comparison. Between strokes, a rest time was taken to wait for the system to become thermally stable and to set constant experimental conditions for each stroke.

## 3. Results

### *3.1. Compression Starting from Atmospheric Pressure*

The gas properties data of compression starting from 1 bar (Compression-1) are given in Figure 3a–d. Figure 3a shows the pressure and temperature data over the entire process. To check the repeatability and consistency of the data collected from each stroke, Figure 3b–d compare the pressure, temperature, and volume of each stroke with a closer look. For these plots, the data from the starting point to around the point of pressure ratio of 2 are displayed.

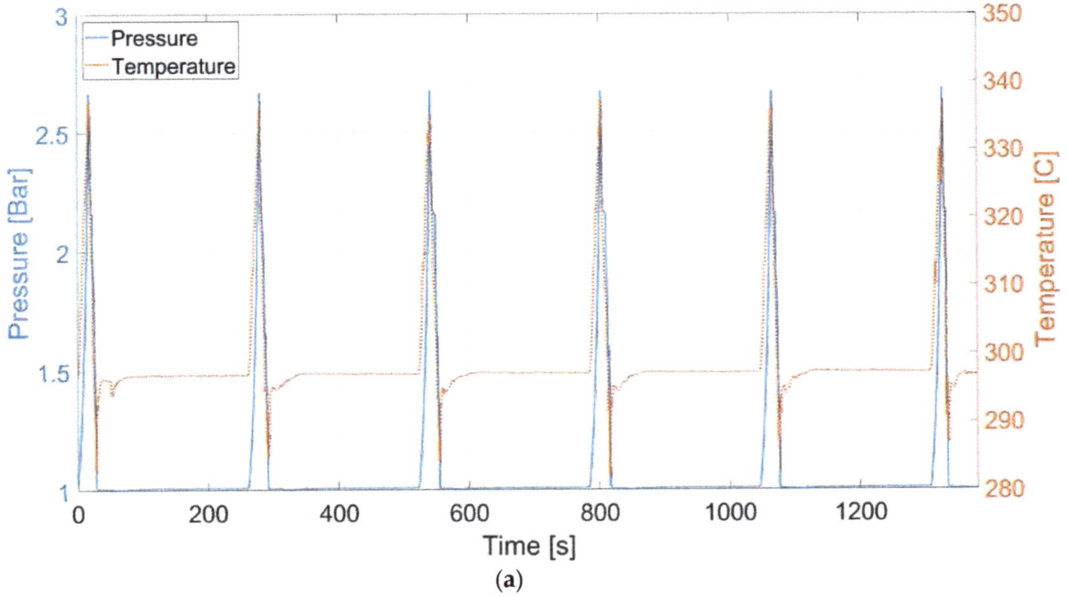


**(a)**

**Figure 3.** *Cont.*

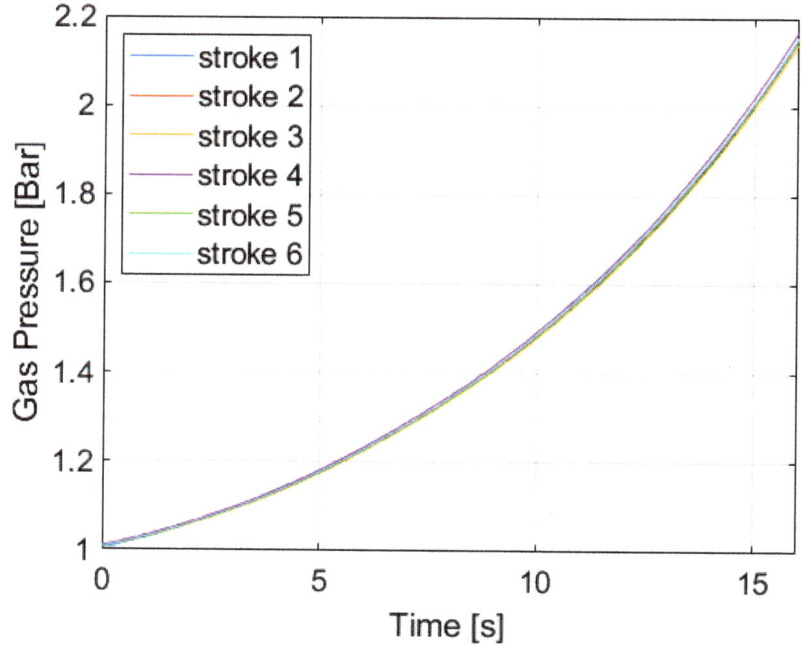


(**b**)

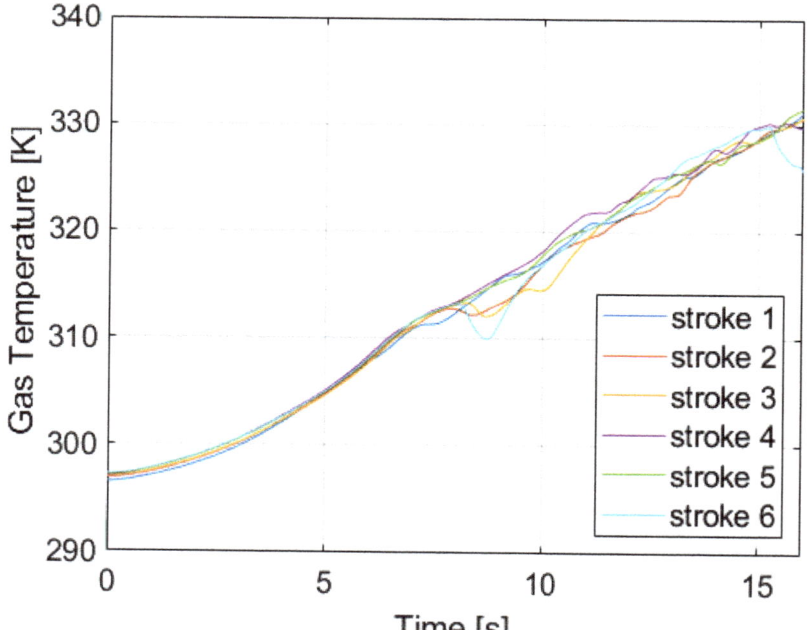


(**c**)

**Figure 3.** *Cont.*

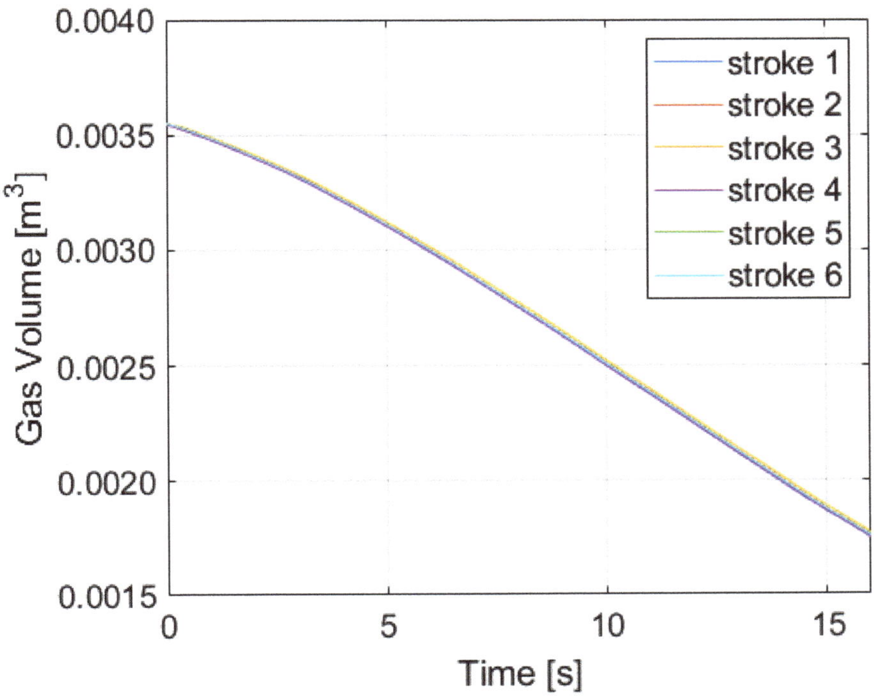


(d)

**Figure 3.** Compression Starting at 1 Bar: (**a**) Pressure and Temperature over Whole Process; (**b**) Pressure, (**c**) Temperature, and (**d**) Volume of Each Stroke.

As can be seen in Figure 3b–d, over the six strokes, the gas properties are similar to one another. This means that the liquid piston system has set-up consistent experimental conditions. In Figure 3c, the gas temperature plot shows oscillation in the values. Because the thermocouple used in the current experiment has a fine wire diameter and an exposed tip to ensure the sensitivity, it can be easily affected by the environment. During compression, the liquid piston movement results in air flow. Because the tip of the thermocouple is exposed to the air flow, it may result in oscillations.

### *3.2. Compression Starting from Elevated Pressure*

For compression starting from 2 bars (Compression-2) and 3 bars (Compression-3), Figure 4 displays the measured properties of the gas in the same manner as Compression-1 data given in Figure 3a. It is observed that each stroke is less consistent if the initial pressure level is higher. This is expected because the water pumping system at higher pressure requires more power to operate and adds instability. However, near the point where a pressure ratio of 2 is achieved, each stroke matches well with one another.

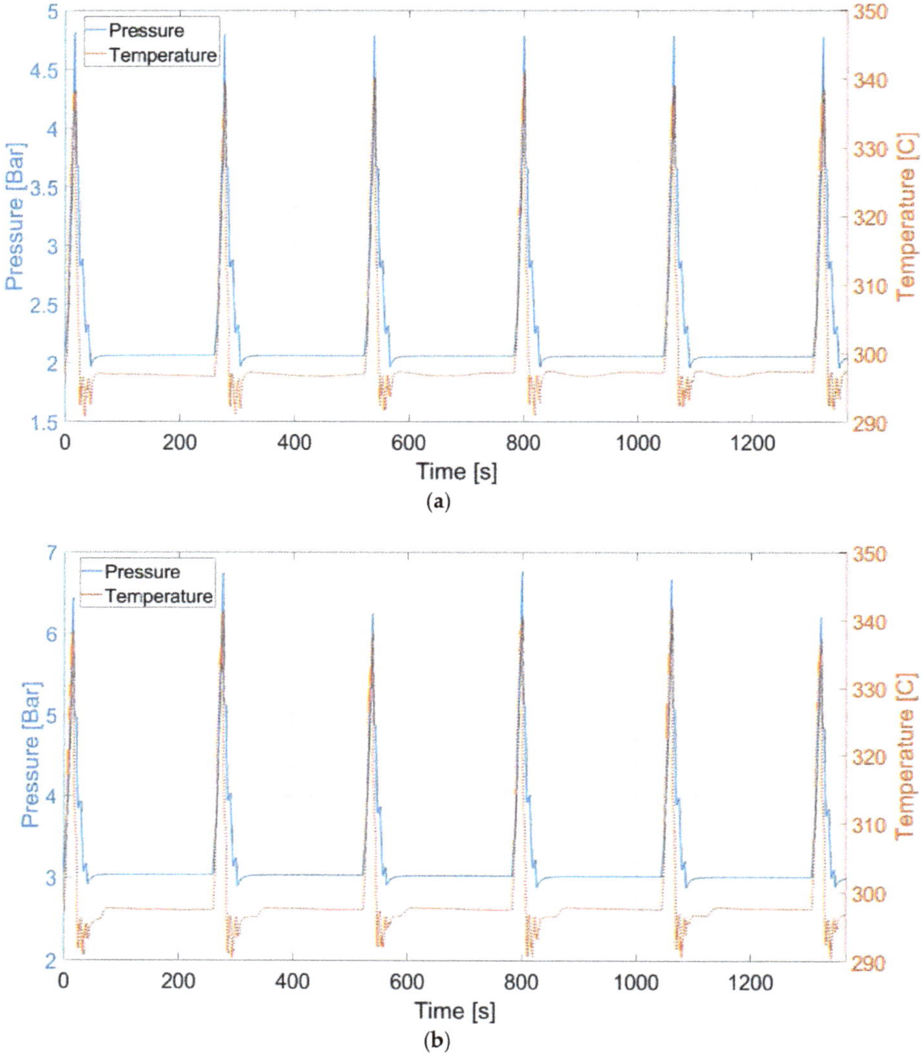


**Figure 4.** Pressure and Temperature over Whole Process: (**a**) Compression Starting at 2 Bar; (**b**) Compression Starting at 3 Bar.

### *3.3. Comparison of Atmospheric and Elevated Pressure Compression*

From the second stroke to the sixth stroke, the five strokes are averaged and the trends of the three cases are compared in Figure 5. The averaged data are collected up to the pressure ratio of 2. C-1, C-2, and C-3 in the legend boxes stand for Compression-1, 2, and 3, respectively. The stroke times have been well matched. To be more specific, for Compression-1 and 3, it takes 15.14 s; for Compression-2, 15.43 s. Figure 6b shows the temperature of the gas during the processes. The temperature increments, $T_f - T_0$, for Compression-1, 2, and 3 are 32.35 K, 38.69 K, and 38.66 K, respectively. Comparing Compression-1 and 2, as the initial pressure is elevated, more temperature increase is observed. However, between Compression-2 and 3, it barely shows a difference. The volume profiles are displayed in Figure 5c. There are differences in the initial volumes because the water level is pushed down after setting up elevated initial pressure, and the differences are reflected on the initial volume calculations.

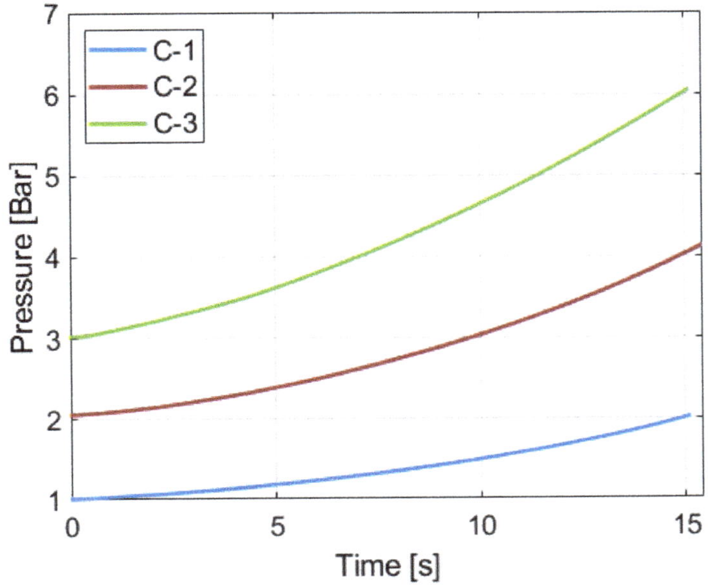


(a)

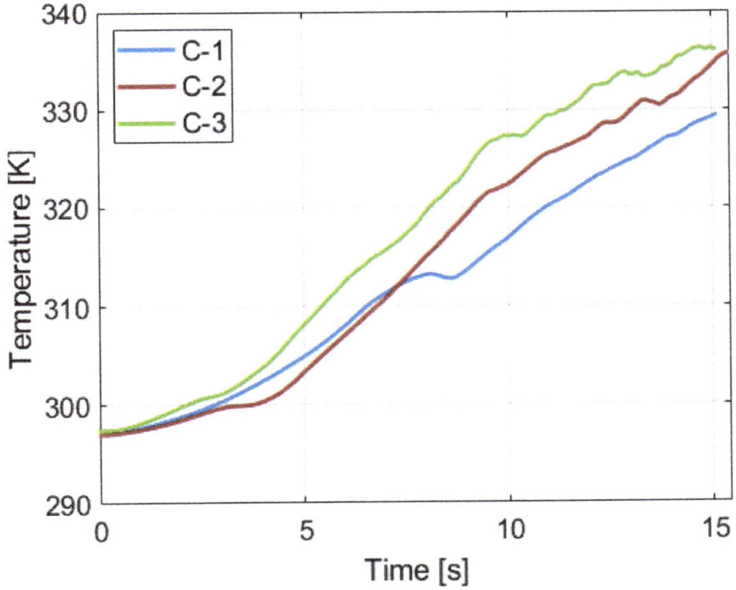


(b)

**Figure 5.** *Cont.*

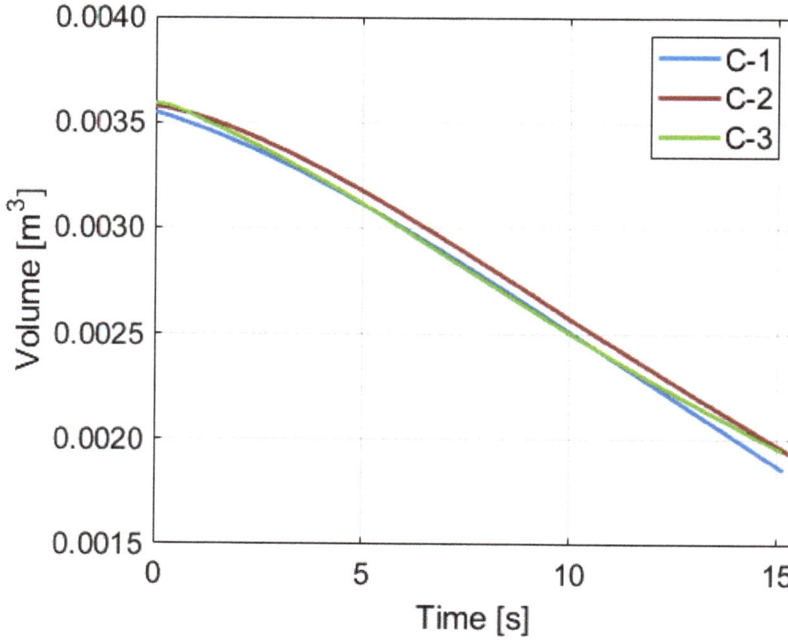


(c)

**Figure 5.** Averaged (**a**) Pressure, (**b**) Temperature, and (**c**) Volume of Compression-1, Compression-2, and Compression-3.

Figure 6a compares the actual pressure–volume experimental data of Compression-1, 2, and 3 to isothermal and adiabatic compression profiles for each initial pressure, and their normalized profiles are displayed in Figure 6b. When the initial pressure is lower, the trajectory is closer to the isothermal trajectory. For at-a-glance comparison, Table 2 presents the starting pressures, temperature increases, and isothermal compression efficiencies. Because the temperature data are local temperatures, it is difficult to determine how close a process is to isothermal compression with the temperature alone. Therefore, by employing Equation (8), isothermal efficiencies of each process are attained for a better comparison.

**Table 2.** Initial Pressure, Temperature Change, and Isothermal Compression Efficiency Comparison of Compression-1, 2, and 3.

| Compression | $P_0$ [Bar] | $T_f - T_0$ [K] | $\eta_{iso}$ [%] |
|---|---|---|---|
| Compression-1 | 1.010 | 32.35 | 89.1 |
| Compression-2 | 2.064 | 38.69 | 85.0 |
| Compression-3 | 3.027 | 38.66 | 83.5 |

Comparing the isothermal efficiencies, the compression performed at a higher pressure is less efficient than that at a lower pressure. The difference between Compression-2 and 3 is smaller than the gap between Compression-1 and 2.

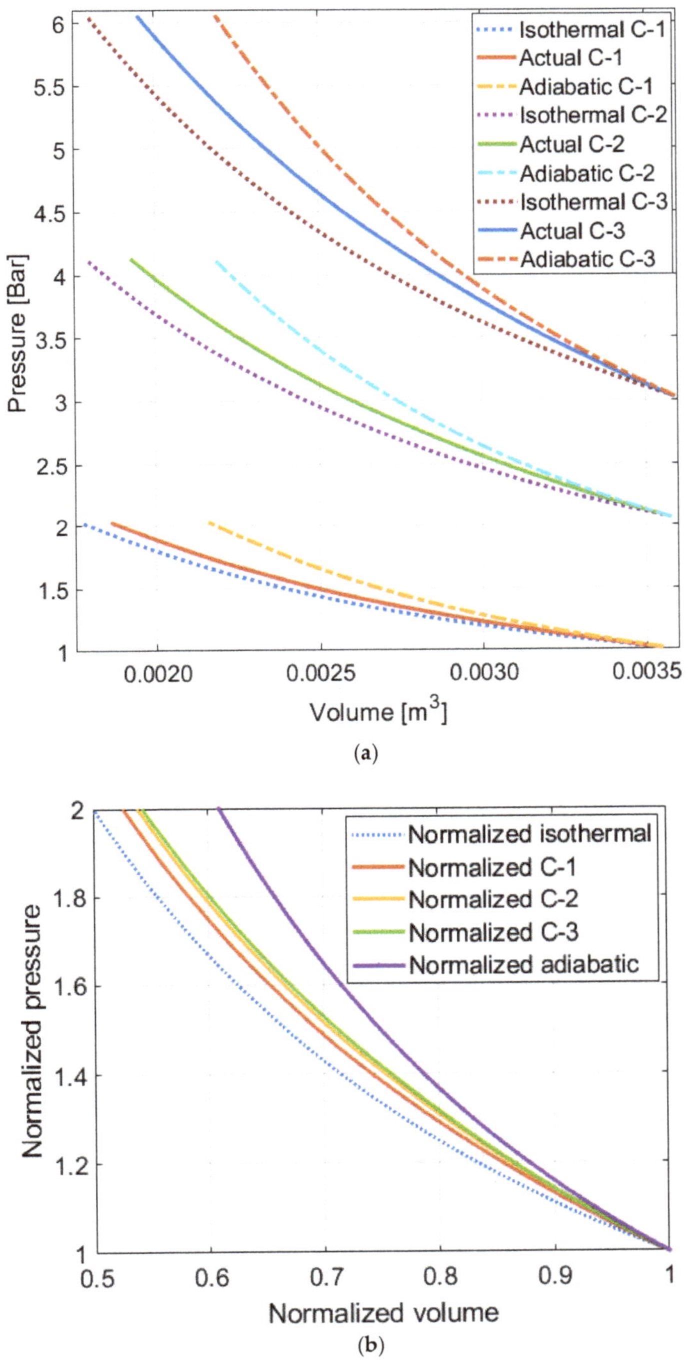


**Figure 6.** Comparison of Actual Process Trajectories with Isothermal and Adiabatic Trajectories: (**a**) Actual Trajectories; (**b**) Normalized Trajectories.

To validate this tendency, an uncertainty analysis is performed considering the data processing and possible error sources of the measurement system. For the efficiency results given in Table 2, the average of five strokes is obtained first. Then, the compression efficiencies are calculated out of the averaged data. To complement the efficiency results, which do not deal with deviation, isothermal efficiencies of each stroke are computed as well. The average and standard deviation of each stroke are presented in Table 3.

**Table 3.** Isothermal Efficiency Comparison of Compression-1, 2, and 3 Calculated from Each Stroke.

| Compression | $\overline{\eta_{iso}}$ | $\sigma_{iso}$ |
|---|---|---|
| Compression-1 | 0.8906 | 0.0013 |
| Compression-2 | 0.8504 | $7.6289 \times 10^{-4}$ |
| Compression-3 | 0.8350 | 0.0014 |

The standard deviations of each process are small, and the lowest value of a lower-pressure case is greater than the highest value of a higher-pressure case. Hence, it shows the tendency that higher pressure leads to lower efficiency. In terms of the system, volume measurement includes some errors. For air volume measurement, the volume of the water entering the chamber is measured by a sensor. Then, the volume of the air can be calculated by subtracting the water volume from the initial air volume inside the chamber. For Compression-2 and 3, when the chamber is pressurized to higher initial pressure settings, the water level is slightly decreased, which makes a slight initial volume change. Although the changes are reflected in the volume calculations, the manual measurement of the initial volume of air includes some possible errors. Nonetheless, the degree of the errors is minor. Furthermore, data processing with and without reflecting the possible errors is made for comparison, and the same efficiency trend is validated.

## 4. Discussion

The results show a trend over the tested cases. The higher the pressure level is, the less efficient the process becomes. When pressure is different, one of the phenomena that occurs is gas dissolution. Therefore, to determine whether the dissolution contributes to the variation in the efficiencies at different pressure levels and to properly evaluate the liquid piston performance, the gas dissolution issue needs to be investigated.

The idea that dissolution takes place during compression can be supported by computing polytropic indices of the compression trajectories. Polytropic indices can be calculated by using Equation (7) and pressure and volume data at two different points. The initial point is used for $P_1$ and $V_1$, and the halfway point of the process for $P_2$ and $V_2$. Figure 7 displays the actual profiles plotted with experimental data and ideal polytropic process profiles from the calculated polytropic indices.

There are mismatches between the actual compression profiles and the ideal polytropic process profiles. Although choosing another point for $P_2$ and $V_2$ can adjust the shape of the profiles, it does not make them perfectly overlap the actual profiles. A change in air mass during the process can cause this issue. One factor that brings about air mass change may be leakage because the system is not perfectly leakage-free. In order to identify how significant the leakage issue is, extra tests are performed. The system performs a compression stroke and holds the pressurized state for some extended period for the initial pressure of 1, 2, and 3 bars. Once it becomes thermally stable after 2 min from the beginning of the process, there is no more notable pressure drop, due to cooling down of the air. After that, the pressure drop per minute is less than 0.01 bar, which is negligibly small for the elevated pressure level achieved. The pressure drop ratio per minute is, at most, 0.15% among the collected data. The tested pressure is even higher than the maximum pressure of interest in the main experiments. For the lower-pressure cases, the pressure drop is even smaller. Provided that a single compression stroke takes less than 20 s in this work, the mass loss during compression due to leakage is negligible. The measured gas properties and plots of the data to support this assumption are summarized in the figures and tables in Appendix A.

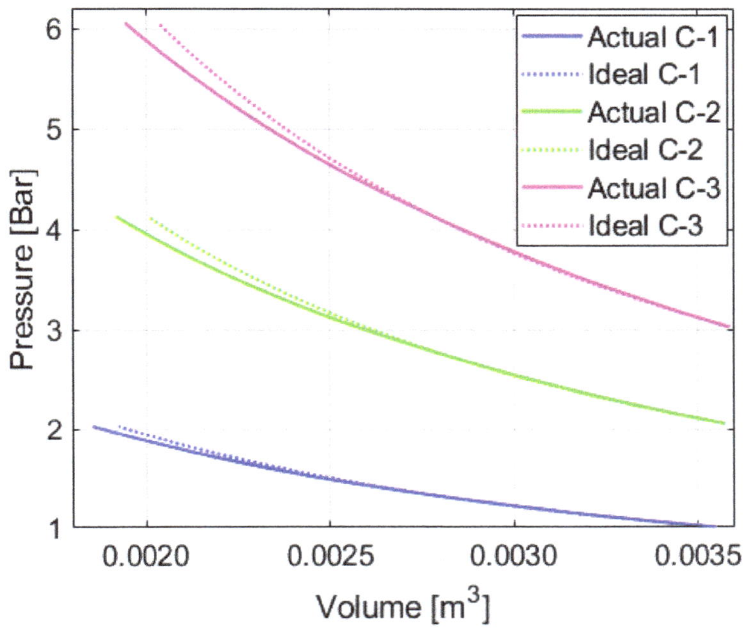


**Figure 7.** Comparison of Actual Profiles and Ideal Polytropic Process Profiles.

Another factor that may cause the mismatch is gas dissolution. Because most of gas dissolution takes place within a very short time in such a liquid-based pressurized system [35], the compression time in this work is long enough for the gas dissolution. Thus, it is reasonable to conclude that the dissolution takes place during the process, and it results in the mismatch between the actual profiles and generated polytropic process profiles. Therefore, although polytropic index is a straightforward indicator that shows how efficient a process is, it should not be used in evaluating the efficiency of compression performed with a liquid piston or a similar system that contains a liquid and a gas.

To determine whether the dissolution is a primary factor that causes the efficiency differences of the processes, the gas solubility changes are estimated. There are two parameters that determine air solubility in water: temperature and pressure. However, Henry's law given in Equation (11) only includes a pressure variable. Therefore, it is difficult to directly employ Equation (11) to estimate how the air mass changes during the process. Thus, simplified estimations are made in the current work with solubility data available on [36] and Henry's law. Two opposing cases are examined based on two different assumptions. First, temperature change is not considered. Owing to the high specific heat of water and the short compression time, the temperature of the water is assumed to be constant throughout compression. Second, the temperature change is counted. With the air pressure and temperature data, the corresponding solubility values at the beginning and the end of the compression can be attained from the solubility table of [36]. For simplification, the measured temperature data are used in finding the corresponding solubility values. Neither of these cases perfectly simulate the actual circumstances of the system during the compression. However, by examining the two extreme cases, the overall trends can be glanced if there is a consistency.

Figure 8 presents the solubilities of air into water at the initial and final points of compression starting from 1, 2, and 3 bars with the pressure ratio of 2. The plot is drawn based upon the data from [36], and how the data are selected is explained in the Appendix A with a supplementary figure. As the data are given in the gauge pressure, Compression-1, 2, and 3 start from 0, 1, and 2 gauge bar and end at 1, 3, and 5 gauge bar, respectively. For the final points, there are two points for each process. The points named const T in the legend box are for the constant-temperature assumption, and the points named gas T are for the second assumption that the system temperature changes the same as the gas temperature. Table 4 summarizes the solubility at the marked points and the amount of increase from the initial points to the final points along with the reduced work input by the dissolution. If the constant temperature is assumed, the solubility increase of Compression-2 is twice as much as Compression-1. For Compression-3, the increase is three times as much as Compression-1. If the temperature is the same as the gas temperature, compared to Compression-1, increases of about 1.91 times for Compression-2 and 3.06 times for Compression-3 are observed, and a lower amount of air is absorbed due to the higher temperature. A similar trend of the proportional increase with the initial pressure is observed under both cases.

**Table 4.** Thermal Properties Comparison of Compression-1, 2, and 3.

| Compression | $S_0$ | $S_{f,const}$ | $S_{f,gas}$ | $W_{dis,const}$ | $W_{dis,gas}$ |
|---|---|---|---|---|---|
| Compression-1 | 0.0177 | 0.035541 | 0.023178 | $892.05V_w$ | $273.90V_w$ |
| Compression-2 | 0.035541 | 0.071223 | 0.046016 | $3568.20V_w$ | $1047.50V_w$ |
| Compression-3 | 0.053382 | 0.106905 | 0.07016 | $8028.45V_w$ | $2516.70V_w$ |

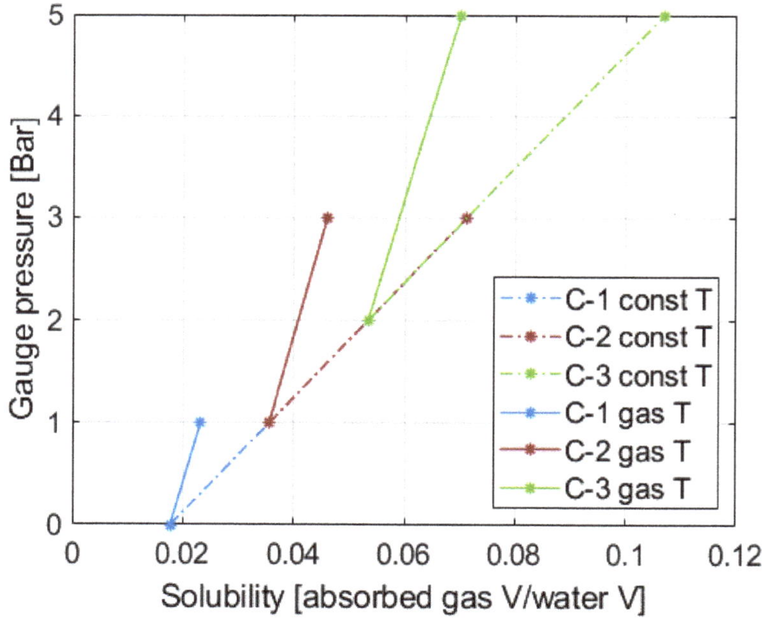


**Figure 8.** Solubility at Initial and Final Points with Two System Temperatures.

Regarding the reduced work input, the higher pressure leads to the greater decrease in work input due to an increased amount of the dissolved gas volume. For both cases, Compression-2 and 3 show roughly 4 times and 9 times the reduced work, compared to Compression-1. Considering that the compression work input of Compression-2 and 3 takes about 2 times and 3 times the work input of Compression-1, the ratios of the reduced work input to the compression work input of Compression-2 and 3 are about 2 and 3 times the ratio of Compression-1, respectively. Despite the greater reduced work of the processes starting from the elevated pressures, Compression-2 and 3 have lower isothermal efficiency than Compression-1. This implies that the gas dissolution is not the primary reason for the different isothermal efficiencies of the present work. Therefore, the results observed in this work of the higher isothermal efficiency of the compression with the lower initial pressure cannot be explained by air dissolution alone. A possible explanation is that air at higher pressure has a greater air mass in a fixed volume. Thus, it may change the motion of air during compression and affect the degree of convection. To accurately identify the full reason for the differences in the efficiency of processes with various initial pressures, microscopic investigation on the dynamics of air inside the chamber should be performed, which is beyond the scope of this work.

The air dissolution complicates the efficiency evaluation with Equation (8). This is because the dissolution compromises the work output. For an energy storage application, the compressed air has to be expanded to restore the energy. When the compressed air is removed from the compressor, the total mass of the compressed air leaving the chamber is less than the original mass before the compression due to the loss caused by dissolution. This will lead to a reduced energy restored in the discharging process. In other words, although the dissolution is beneficial for the isothermal efficiency on paper obtained using Equation (8), the actual efficiency based on how much work can be restored is different from the calculated value. Because each compression process has different dissolved amounts, the compromised work output may vary as well. To evaluate how much efficiency is compromised, the output work needs to be quantified. Because of the difficulty of predicting the exact expansion trajectory without empirical work, the volume of air after the expansion is quantified and compared by estimating the mass changes of the air. The ideal gas equation and the solubility data from [36] enable the volume quantification. The absorbed gas volume varies with the water volume.

Figure 9 displays the mass of the air that is not dissolved into the water and remains as a gas above the water. This estimation needs water volume in a system and air solubility. For water volume, the volume is assumed to be the same as the initial air volume in the chamber. For solubility changes, the solubilities under the constant-temperature assumption are used. To perform a proportional comparison of the three processes, the mass and pressure values are normalized by dividing by the initial values. At the target pressure, the ratios of the mass to the initial mass for Compression-1, 2, and 3 are 94.57%, 88.96%, and 83.81%, respectively. After expansion up to the initial pressure of the pre-compression status, the restorable volume will also be 94.57%, 88.96%, and 83.81% of the initial volume, respectively. This means that an increased solubility at the higher pressure results in a greater compromise in the work output unless the expansion takes place in the same chamber and the dissolved air is fully restored.

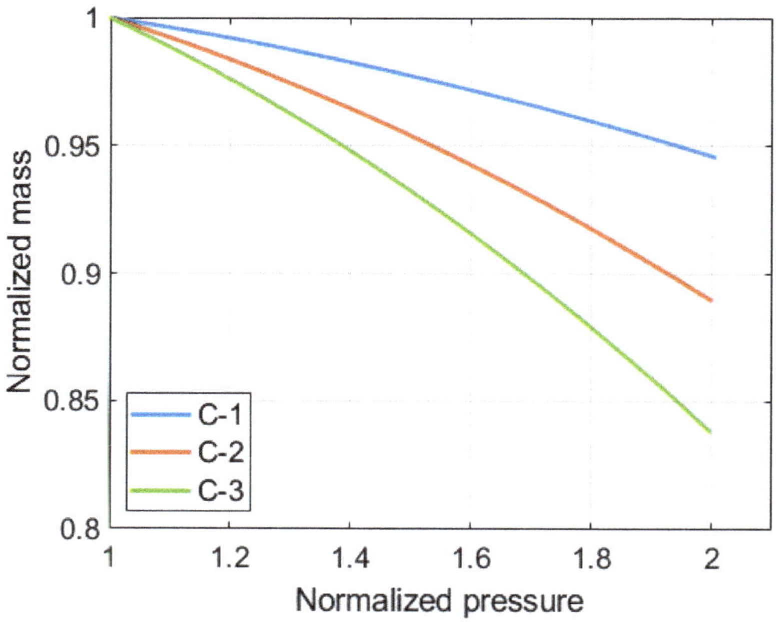


**Figure 9.** Comparison of Normalized Air Mass—Pressure Changes of Compression-1, 2, 3.

Separately, a significant change in the water volume also has an impact on the soluble amount. Table 5 presents the variations in the air mass ratios with different water volumes.

**Table 5.** Mass Ratio after Compression with Different Water Volumes.

| Water Volume | Air Mass Ratio to Initial Air Mass [%] | | |
|---|---|---|---|
| | Compression-1 | Compression-2 | Compression-3 |
| $0.5V_{0,gas}$ | 97.28 | 94.48 | 91.90 |
| $V_{0,gas}$ | 94.57 | 88.96 | 83.81 |
| $1.5V_{0,gas}$ | 91.85 | 83.43 | 75.71 |

When the water volume is smaller, the absolute amount of the dissolved air mass is smaller, leading to a larger air mass ratio. In contrast, when the water volume is larger, the ratios become smaller. Furthermore, the differences in the ratios between the processes become greater, which results in a significantly lower value for Compression-3 when the water volume is 1.5 times the initial air volume. Hence, using the minimum amount of water will help secure the restorable work output. Although those numbers are generated based on a simple simulation model, it implies how the volume of water can affect the system efficiency.

The dissolution is advantageous for the isothermal compression efficiency on paper calculated using Equation (8) because it reduces the volume of the air inside the chamber as the pressure increases. Thus, the process with higher initial pressure benefits more from an increased dissolution. However, despite a greater dissolution, the higher pressure shows the lower efficiency. On top of the lower efficiency at the higher pressure, the greater dissolution will lead to a lower restoration of the output energy. Hence, for the overall system including the energy discharging stage considered, there will be an even larger gap in the actual efficiencies than their calculated values. Although the higher-pressure process has the advantage of having a higher power to perform the charging/discharging process faster, it shows compromised performance with regard to overall efficiency. However, this should not depreciate the efficacy of the liquid piston compressor. The main strength of the liquid piston compressor is that various heat transfer enhancement techniques can be easily integrated regardless of the dissolution.

There are limitations to the current study. As referred earlier, the overall temperature of the air and the dissolved air mass cannot be obtained with the ideal gas equation, because each one requires the value of the other one. As the simplified estimation of the total solubility changes is used for the amount of the dissolved air, errors from the simplification may be inevitable. In addition, there is room for improvement in efficiency analysis. The efficiency, $\eta_{iso}$, is calculated by comparing the isothermal work, $W_{isothermal}$, and the actual work, $W_{actual}$. In the current study, the mass change is not considered in calculating $W_{isothermal}$, while $W_{actual}$ is calculated from the data collected from the real experiment where dissolution takes place. If the isothermal profile reflects the mass change as well, it could possibly lead to a different conclusion.

Nevertheless, to the author's best knowledge, this is the first work to experimentally investigate how the initial pressure level affects the isothermal compression efficiency when the pressure ratio is the same and to determine whether the gas dissolution is associated with the different compression efficiencies. Furthermore, by testing compression at different initial pressures, an initial attempt is made to simulate the applicability of the liquid piston compressor to a multi-stage compression system. The examined pressure ranges are quite limited to lower pressures because Henry's law is not applicable if the pressure goes up high enough [40]. At higher pressures, the trend may be different from the evaluated range of the current work. Compression at much higher levels to study how the trend changes will be an interesting topic for future research.

## 5. Conclusions

The impacts of different initial pressures on liquid piston compression are investigated in this paper by performing compressions starting from 1 bar, 2 bars, and 3 bars with the same pressure ratio of 2. At elevated pressures, the processes result in higher temperature increases over the compression that starts from the atmospheric pressure, whereas there is no notable difference in temperature rise between compressions starting at 2 bars and 3 bars. Regarding compression efficiency, the compressions at higher pressure lead to lower isothermal efficiencies. Experimental results have shown that the isothermal efficiencies of Compression-1, 2, and 3 processes are 89.1%, 85.0%, and 83.5%, respectively. To identify the cause of the efficiency differentiation, gas solubility changes of the three compression processes are quantified based on simple estimation models. Because the gas solubility is proportional to the pressure, the higher initial pressure process has more gas absorbed in the water. Despite the increased solubility, which is beneficial for isothermal efficiency calculations on paper, the compression processes at higher pressures have shown lower efficiencies than the compression starting from the atmospheric pressure. Hence, the dissolution is not the primary reason for the lower compression efficiency at higher initial pressure. Because the air behaves differently during compression depending on the pressure and the mass in the chamber, a comprehensive analysis on the dynamics of the air in the chamber may help clarify the cause of the phenomenon.

Although the dissolution is advantageous for the isothermal efficiency calculations on paper, the dissolved mass may further compromise the overall system efficiency than the calculated values. This is because the isothermal efficiency can be calculated in a conventional manner without considering air mass changes caused by the dissolution. For a real application that includes the energy discharging process, dissolved mass results in a reduced volume of air after expansion and it leads to less energy output. Assuming the system contains the same water volume as the initial air volume, the masses of the air after compressions starting at 1, 2, and 3 bars are estimated to be 94.57%, 88.96%, and 83.81% of the initial masses prior to the compression, respectively, which further compromises the system efficiency than the calculated compression efficiencies discounting dissolution. In conclusion, even though the dissolution is not a major factor that results in efficiency decrease at a higher pressure level on paper, it actually further degrades the overall system efficiency.

**Author Contributions:** System design, B.A.; investigation, B.A.; formal analysis, B.A.; writing—original draft, B.A.; conceptualization, P.I.R.; supervision, P.I.R.; writing—review and editing, P.I.R. All authors have read and agreed to the published version of the manuscript.

**Funding:** This research received no external funding.

**Data Availability Statement:** Data available on request due to privacy.

**Conflicts of Interest:** The authors declare no conflict of interest.

## Nomenclature

| **Nomenclature** | |
|---|---|
| $A_s$ | Surface area |
| $C_v$ | Specific heat |
| $k_H$ | Proportionality constant |
| $m$ | Mass |
| $m_{gas}$ | Gas mass |
| $n$ | Polytropic index |
| $P$ | Pressure |
| $P_0$ | Initial pressure |
| $P_1$ | Pressure at point 1 |
| $P_2$ | Pressure at point 2 |
| $P_f$ | Final pressure |
| $P_{gas}$ | Gas pressure |
| $Pc_{min}$ | Pressure change for a minute |
| $Pr$ | Pressure ratio |
| $Pr_{min}$ | Pressure ratio for a minute |
| $\dot{Q}$ | Heat transfer rate |
| $R$ | Gas constant |
| $S_0$ | Initial gas solubility |
| $S_f$ | Final gas solubility |
| $S_{f,const}$ | Final gas solubility with constant-temperature assumption |
| $S_{f,gas}$ | Final gas solubility with gas temperature assumption |
| $S_{gas}$ | Gas solubility |
| $T$ | Temperature |
| $T_0$ | Initial temperature |
| $T_f$ | Final temperature |
| $T_{gas}$ | Gas temperature |
| $T_i$ | Temperature at current step |
| $T_{i-1}$ | Temperature at previous step |
| $T_\infty$ | Temperature of surroundings |
| $t$ | Time |
| $\dot{U}$ | Internal energy rate |
| $U_h$ | Overall heat transfer coefficient |
| $V$ | Volume |
| $V_0$ | Initial volume |
| $V_{0,gas}$ | Initial gas volume |
| $V_1$ | Volume at point 1 |
| $V_2$ | Volume at point 2 |
| $V_f$ | Final volume |
| $V_{f,iso}$ | Final volume of isothermal compression |

| | |
|---|---|
| $V_{gas}$ | Gas volume |
| $V_i$ | Volume at current step |
| $V_{i-1}$ | Volume at previous step |
| $V_w$ | Volume of water |
| $W_{actual}$ | Actual work |
| $W_{comp}$ | Compression work |
| $W_{cooling}$ | Cooling work |
| $W_{dis}$ | Work from dissolution |
| $W_{dis,const}$ | Dissolution work with constant-temperature assumption |
| $W_{dis,gas}$ | Dissolution with gas temperature assumption |
| $W_{isothermal}$ | Isothermal compression work |
| $\dot{W}_{comp}$ | Compression work rate |
| **Greek letters** | |
| $\eta_{iso}$ | Isothermal compression efficiency |
| $\overline{\eta_{iso}}$ | Average isothermal compression efficiency |
| $\sigma_{iso}$ | Standard deviation of isothermal efficiency |
| **Abbreviation** | |
| CAES | Compressed air energy storage |
| A-CAES | Adiabatic compressed air energy storage |
| D-CAES | Diabatic compressed air energy storage |
| I-CAES | Isothermal compressed air energy storage |
| Compression-1 | Compression starting from 1 bar |
| Compression-2 | Compression starting from 2 bar |
| Compression-3 | Compression starting from 3 bar |
| C-1 | Compression-1 (used in plot legend) |
| C-2 | Compression-2 (used in plot legend) |
| C-3 | Compression-3 (used in plot legend) |
| Hold-1 | Hold process starting from 1 bar |
| Hold-2 | Hold process starting from 2 bar |
| Hold-3 | Hold process starting from 3 bar |

## Appendix A

Figure A1 and Table A1 are supplementary materials to substantiate the fact that the leakage of the system is negligible. Compressions are performed from the initial pressure of 1, 2, and 3 bars. Then, without performing expansion, the pressurized state is held for an extended time stretch. The pressure drops of the three cases per minute are summarized in Table A1. The percentages of the current pressure to the pressure measured a minute ago are given in the same table. It takes about two minutes to be nearly thermally stable after compression starts. From the point, the pressure drop per minute does not exceed 0.01 bar at any point. In terms of pressure drop percentage, 99.85% is the smallest value. Overall, the pressure drop is smaller when the pressure is lower. The held pressures for the leakage test are higher than the pressure of interest of the main experiment. Furthermore, the stroke time of the main experiment takes less than 20 s. Thus, the pressure drops that occur in the main tests are expected to be smaller than those values. Thus, the mass losses during the test due to leakage are negligible.

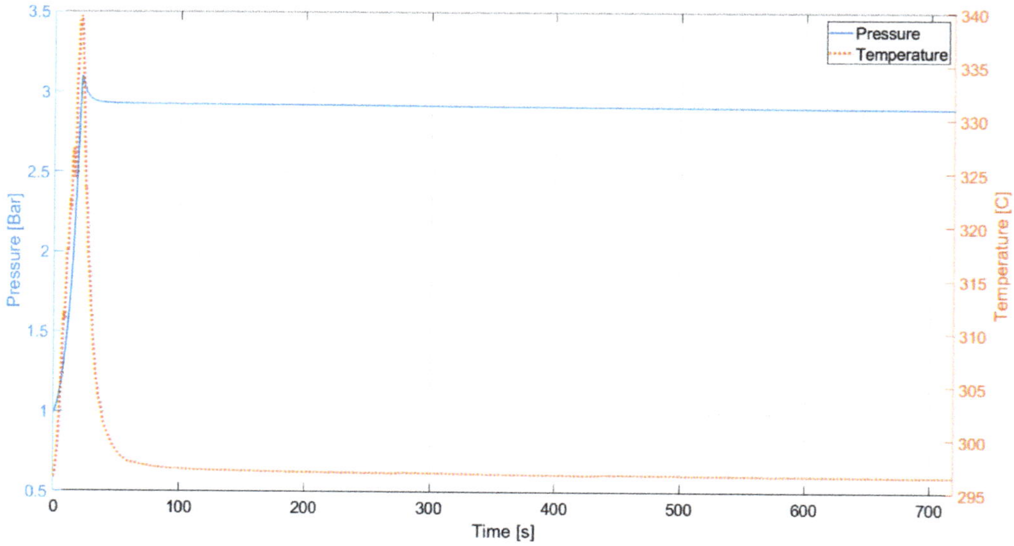


(**a**)

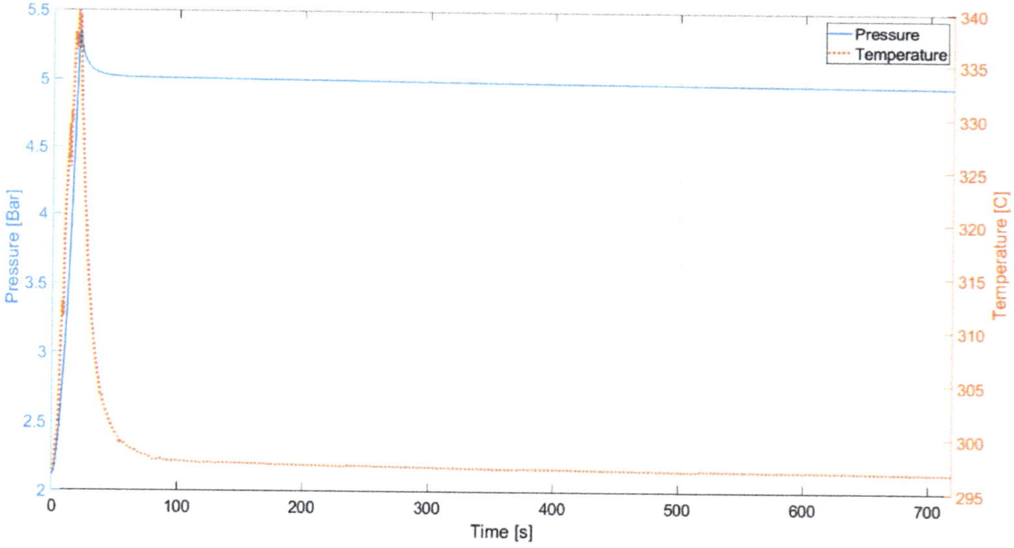


(**b**)

**Figure A1.** *Cont.*

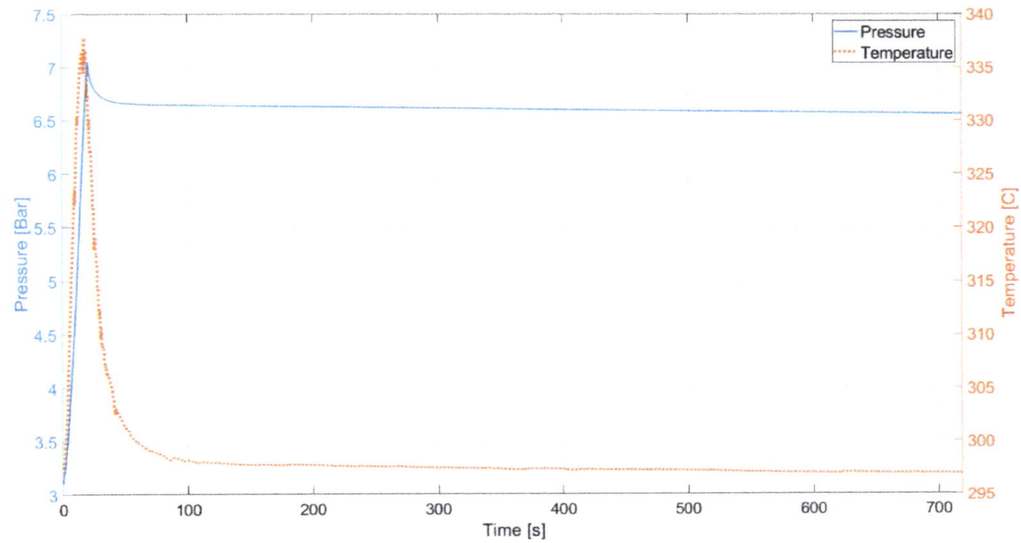


(c)

**Figure A1.** Pressure and Measured Temperature of Compression-and-Hold Process Starting from (**a**) 1 Bar, (**b**) 2 Bar, and (**c**) 3 Bar for Extended Time Stretch (12 Minutes).

**Table A1.** Pressure Data for Hold-1, 2, and 3.

| | **PRESSURE** | | | | | | | | |
|---|---|---|---|---|---|---|---|---|---|
| | **Hold-1** | | | **Hold-2** | | | **Hold-3** | | |
| **Time [s]** | **[Bar]** | $Pc_{minute}$ | $Pr_{minute}$ | **[Bar]** | $Pc_{minute}$ | $Pr_{minute}$ | **[Bar]** | $Pc_{minute}$ | $Pr_{minute}$ |
| 60 | 2.928894 | | | 5.017834 | | | 6.656374 | | |
| 120 | 2.924585 | −0.00431 | 99.85% | 5.008354 | −0.00948 | 99.81% | 6.641895 | −0.01448 | 99.78% |
| 180 | 2.922689 | −0.0019 | 99.94% | 5.000769 | −0.00758 | 99.85% | 6.633276 | −0.00862 | 99.87% |
| 240 | 2.920103 | −0.00259 | 99.91% | 4.995254 | −0.00552 | 99.89% | 6.625347 | −0.00793 | 99.88% |
| 300 | 2.917173 | −0.00293 | 99.90% | 4.988186 | −0.00707 | 99.86% | 6.617591 | −0.00776 | 99.88% |
| 360 | 2.915449 | −0.00172 | 99.94% | 4.982154 | −0.00603 | 99.88% | 6.609145 | −0.00845 | 99.87% |
| 420 | 2.912347 | −0.0031 | 99.89% | 4.976638 | −0.00552 | 99.89% | 6.599664 | −0.00948 | 99.86% |
| 480 | 2.910106 | −0.00224 | 99.92% | 4.970777 | −0.00586 | 99.88% | 6.593976 | −0.00569 | 99.91% |
| 540 | 2.907176 | −0.00293 | 99.90% | 4.964744 | −0.00603 | 99.88% | 6.586047 | −0.00793 | 99.88% |
| 600 | 2.90459 | −0.00259 | 99.91% | 4.959573 | −0.00517 | 99.90% | 6.578118 | −0.00793 | 99.88% |
| 660 | 2.902866 | −0.00172 | 99.94% | 4.955092 | −0.00448 | 99.91% | 6.571051 | −0.00707 | 99.89% |
| 720 | 2.899936 | −0.00293 | 99.90% | 4.949748 | −0.00534 | 99.89% | 6.564846 | −0.00621 | 99.91% |

Figure A2 is a supplementary material to clarify how the solubility estimation was made. The asterisk marks are the data points of [36], and the lines are drawn connecting the points at the lowest and the highest pressure points to estimate the solubility values between the data points. The initial points and final points with the two different assumptions of Compression-1, 2, and 3 are marked on the lines. In Figure A2, (i) stands for the initial point, (f,c) for the final point with the constant-temperature assumption, and (f,g) for the final point with the changing-temperature assumption. Because the dataset does not have exactly same values as the experimental data, the close values that represent the temperature changes well are chosen. The measured and chosen temperature values are presented in Table A2.

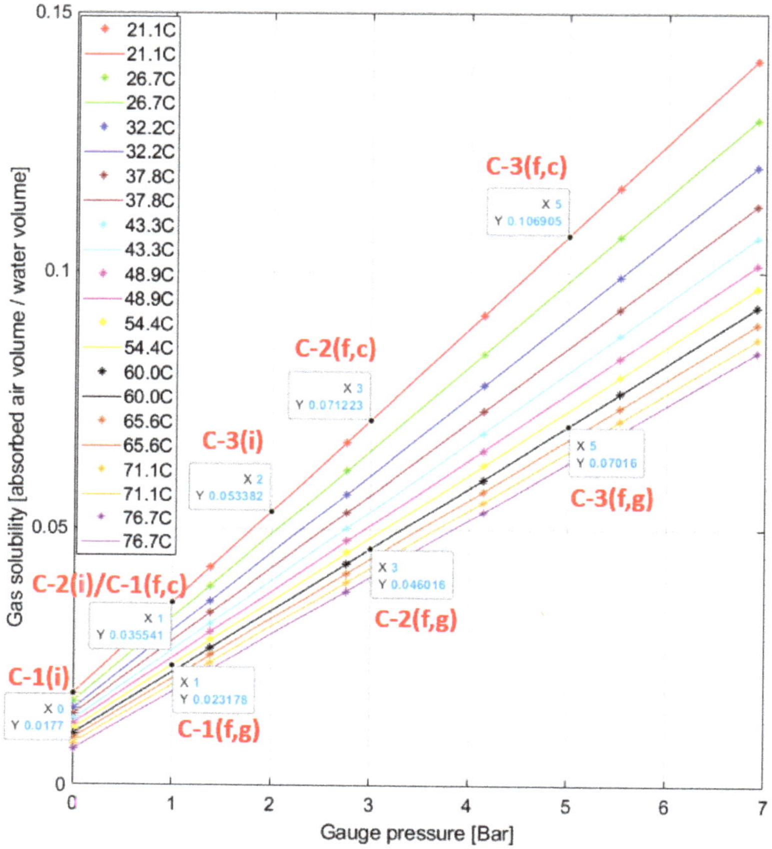


**Figure A2.** Solubility Data.

**Table A2.** Pressure Data for Hold-1, 2, and 3.

| Compression | Measured $T_0$ [°C] | Chosen $T_0$ [°C] | Measured $T_f$ [°C] | Chosen $T_f$ [°C] |
|---|---|---|---|---|
| Compression-1 | 23.9 | 21.1 | 56.3 | 54.4 |
| Compression-2 | 23.9 | 21.1 | 62.6 | 60.0 |
| Compression-3 | 24.3 | 21.1 | 63.0 | 60.0 |

As the temperature data are collected by a thermocouple installed at a specific location, they are local temperature and do not represent the overall temperature of the system. Because of the air mass change caused by dissolution, the air temperature cannot be calculated by the ideal gas equation. Thus, assuming the experimental temperature data represent the overall system temperature, the solubilities are estimated.

For the first assumption that temperature change is not considered and pressure is the only variable, it has a linear profile. On the contrary, for the second assumption that the gas temperature is considered, it is highly difficult to accurately determine the solubility changing profiles throughout the compression. Thus, for the problem simplification, it is assumed that the solubility differences between the two points are the total solubility changes, and the solubility changes linearly with pressure between the two points.

## References


1. Mertz, O.; Halsnæs, K.; Olesen, J.E.; Rasmussen, K. Adaptation to Climate Change in Developing Countries. *Environ. Manag.* **2009**, *43*, 743–752. [CrossRef] [PubMed]
2. Renewable Energy Explained–U.S. Energy Information Administration (EIA). Available online: https://www.eia.gov/energyexplained/renewable-sources/ (accessed on 4 October 2022).
3. Evans, A.; Strezov, V.; Evans, T.J. Assessment of Utility Energy Storage Options for Increased Renewable Energy Penetration. *Renew. Sustain. Energy Rev.* **2012**, *16*, 4141–4147. [CrossRef]
4. Park, J.-K.; Ro, P.I.; He, X.; Mazzoleni, A.P. Analysis and Proof-of-Concept Experiment of Liquid-Piston Compression for Ocean Compressed Air Energy Storage (OCAES) System. In Proceedings of the 2nd Marine Energy Technology Symposium, Seattle, WA, USA, 15 April 2014.
5. Cavallo, A.J. Energy Storage Technologies for Utility Scale Intermittent Renewable Energy Systems. *J. Sol. Energy Eng.* **2001**, *123*, 387–389. [CrossRef]
6. Cavallo, A. Controllable and Affordable Utility-Scale Electricity from Intermittent Wind Resources and Compressed Air Energy Storage (CAES). *Energy* **2007**, *32*, 120–127. [CrossRef]
7. Crotogino, F.; Mohmeyer, K.-U.; Scharf, R. Huntorf CAES: More than 20 Years of Successful Operation. In Proceedings of the SMRI Spring Meeting, Orlando, FL, USA, 23–25 April 2001; Volume 2001.
8. Caralis, G.; Christakopoulos, T.; Karellas, S.; Gao, Z. Analysis of Energy Storage Systems to Exploit Wind Energy Curtailment in Crete. *Renew. Sustain. Energy Rev.* **2019**, *103*, 122–139. [CrossRef]
9. Kim, Y.-M.; Lee, J.-H.; Kim, S.-J.; Favrat, D. Potential and Evolution of Compressed Air Energy Storage: Energy and Exergy Analyses. *Entropy* **2012**, *14*, 1501–1521. [CrossRef]
10. Patil, V.C.; Ro, P.I. Energy and Exergy Analysis of Ocean Compressed Air Energy Storage Concepts. *J. Eng.* **2018**, *2018*, e5254102. [CrossRef]
11. Budt, M.; Wolf, D.; Span, R.; Yan, J. Compressed Air Energy Storage–An Option for Medium to Large Scale Electrical-Energy Storage. *Energy Procedia* **2016**, *88*, 698–702. [CrossRef]
12. Van de Ven, J.D.; Li, P.Y. Liquid Piston Gas Compression. *Appl. Energy* **2009**, *86*, 2183–2191. [CrossRef]
13. Qin, C.; Loth, E. Liquid Piston Compression Efficiency with Droplet Heat Transfer. *Appl. Energy* **2014**, *114*, 539–550. [CrossRef]
14. Qin, C.; Loth, E.; Li, P.; Simon, T.; Van de Ven, J. Spray-Cooling Concept for Wind-Based Compressed Air Energy Storage. *J. Renew. Sustain. Energy* **2014**, *6*, 043125. [CrossRef]
15. Electricity Generation from Wind–U.S. Energy Information Administration (EIA). Available online: https://www.eia.gov/energyexplained/wind/electricity-generation-from-wind.php (accessed on 4 February 2023).
16. Wind Energy. Available online: https://www.irena.org/Energy-Transition/Technology/Wind-energy (accessed on 4 February 2023).
17. Zhang, X.; Xu, Y.; Zhou, X.; Zhang, Y.; Li, W.; Zuo, Z.; Guo, H.; Huang, Y.; Chen, H. A Near-Isothermal Expander for Isothermal Compressed Air Energy Storage System. *Appl. Energy* **2018**, *225*, 955–964. [CrossRef]
18. Yan, B.; Wieberdink, J.; Shirazi, F.; Li, P.Y.; Simon, T.W.; Van de Ven, J.D. Experimental Study of Heat Transfer Enhancement in a Liquid Piston Compressor/Expander Using Porous Media Inserts. *Appl. Energy* **2015**, *154*, 40–50. [CrossRef]
19. Wieberdink, J.; Li, P.Y.; Simon, T.W.; Van de Ven, J.D. Effects of Porous Media Insert on the Efficiency and Power Density of a High Pressure (210 bar) Liquid Piston Air Compressor/Expander–An Experimental Study. *Appl. Energy* **2018**, *212*, 1025–1037. [CrossRef]
20. Patil, V.C.; Liu, J.; Ro, P.I. Efficiency Improvement of Liquid Piston Compressor Using Metal Wire Mesh for Near-Isothermal Compressed Air Energy Storage Application. *J. Energy Storage* **2020**, *28*, 101226. [CrossRef]
21. Khaljani, M.; Vennard, A.; Harrison, J.; Surplus, D.; Murphy, A.; Mahmoudi, Y. Experimental and Modelling Analysis of Efficiency Enhancement in a Liquid Piston Gas Compressor Using Metal Plate Inserts for Compressed Air Energy Storage Application. *J. Energy Storage* **2021**, *43*, 103240. [CrossRef]
22. Patil, V.C.; Acharya, P.; Ro, P.I. Experimental Investigation of Water Spray Injection in Liquid Piston for Near-Isothermal Compression. *Appl. Energy* **2020**, *259*, 114182. [CrossRef]
23. Ahn, B.; Patil, V.C.; Ro, P.I. Effect of Integrating Metal Wire Mesh with Spray Injection for Liquid Piston Gas Compression. *Energies* **2021**, *14*, 3723. [CrossRef]
24. Patil, V.C.; Ro, P.I. Experimental Study of Heat Transfer Enhancement in Liquid Piston Compressor Using Aqueous Foam. *Appl. Therm. Eng.* **2020**, *164*, 114441. [CrossRef]
25. Ren, T.; Xu, W.; Cai, M.; Wang, X.; Li, M. Experiments on Air Compression with an Isothermal Piston for Energy Storage. *Energies* **2019**, *12*, 3730. [CrossRef]
26. Ren, T.; Xu, W.; Jia, G.-W.; Cai, M. A Novel Isothermal Compression Method for Energy Conservation in Fluid Power Systems. *Entropy* **2020**, *22*, 1015. [CrossRef] [PubMed]
27. Li, C.; Wang, H.; He, X.; Zhang, Y. Experimental and Thermodynamic Investigation on Isothermal Performance of Large-Scaled Liquid Piston. *Energy* **2022**, *249*, 123731. [CrossRef]

28. Zhang, C.; Li, P.Y.; Van de Ven, J.D.; Simon, T.W. Design Analysis of a Liquid-Piston Compression Chamber with Application to Compressed Air Energy Storage. *Appl. Therm. Eng.* **2016**, *101*, 704–709. [CrossRef]
29. Patil, V.C.; Acharya, P.; Ro, P.I. Experimental Investigation of Heat Transfer in Liquid Piston Compressor. *Appl. Therm. Eng.* **2019**, *146*, 169–179. [CrossRef]
30. Hu, S.; Xu, W.; Cai, M.; Jia, G. Energy Efficiency and Power Density Analysis of a Tube Array Liquid Piston Air Compressor/Expander for Compressed Air Energy Storage. *J. Energy Storage* **2022**, *55*, 105674. [CrossRef]
31. Single, vs. Multi-Stage Compressors. Available online: https://cascousa.com/compressed-air-101/types-of-compressors/single-versus-multi-stage-compressors/ (accessed on 7 April 2022).
32. Budt, M.; Wolf, D.; Span, R.; Yan, J. A Review on Compressed Air Energy Storage: Basic Principles, Past Milestones and Recent Developments. *Appl. Energy* **2016**, *170*, 250–268. [CrossRef]
33. Wolf, D. Methods for Design and Application of Adiabatic Compressed Air Energy Storage Based on Dynamic Modeling. 2011. Available online: https://www.osti.gov/etdeweb/biblio/21461477 (accessed on 4 February 2023).
34. Huang, L.; Simon, T.W.; Li, P.Y.; Hummelt, E.J. Analysis of Dissolved Gas in the Application of Liquid Piston Gas Compression. In Proceedings of the 1st Thermal and Fluid Engineering Summer Conference, TFESC, New York, NY, USA, 9–12 August 2015; pp. 9–12.
35. Xu, W.; Garvey, S.D.; Ren, T.; Hu, Y. Dynamics of Dissolution for Underwater Compressed Air Energy Storage. In Proceedings of the 2019 Offshore Energy and Storage Summit (OSES), Brest, France, 10–12 July 2019; pp. 1–5.
36. Engineering ToolBox (2004) Solubility of Air in Water. Available online: https://www.engineeringtoolbox.com/air-solubility-water-d_639.html (accessed on 11 November 2022).
37. Çengel, Y.; Boles, M. *Thermodynamics An Engineering Approach*, 2nd ed.; McGraw-Hill, Inc.: Hightstown, NJ, USA, 1994; ISBN 0-07-911652-3.
38. Zhang, C.; Yan, B.; Wieberdink, J.; Li, P.Y.; Van de Ven, J.D.; Loth, E.; Simon, T.W. Thermal Analysis of a Compressor for Application to Compressed Air Energy Storage. *Appl. Therm. Eng.* **2014**, *73*, 1402–1411. [CrossRef]
39. Bhaskaran, H.S.; Ro, P.; Park, J.-K.; Ramakrishnan, K.R. Analysis of a Novel Technique for Temperature Rise Abatement in Liquid Piston Compressors–External Gas Injection. *J. Therm. Sci. Eng. Appl.* **2017**, *9*, 024503. [CrossRef]
40. Henry's Law. Available online: https://chem.libretexts.org/Bookshelves/Physical_and_Theoretical_Chemistry_Textbook_Maps/Supplemental_Modules_(Physical_and_Theoretical_Chemistry)/Physical_Properties_of_Matter/Solutions_and_Mixtures/Ideal_Solutions/Dissolving_Gases_In_Liquids%2C_Henry's_Law (accessed on 5 October 2022).

energies

MDPI

*Article*

# Estimating the Drivers of the Cost of Saved Electricity in Utility Customer-Funded Energy Efficiency Programs

**Sean Murphy * and Natalie Mims Frick**

Electricity Markets and Policy Department, Lawrence Berkeley National Laboratory, 1 Cyclotron Road, Berkeley, CA 94720, USA
* Correspondence: smurphy@lbl.gov

**Abstract:** Energy efficiency programs funded by utility customers provide an electricity resource in most U.S. states, but their scale and cost of saving electricity varies significantly by state. In this paper, we explore the drivers of the cost of saved electricity in these programs with an econometric model and nearly a decade of data reported by efficiency program administrators. We found strong evidence for economies of scale and weak evidence for diseconomies of scale, which suggests that states with low levels of efficiency savings relative to retail sales can increase the size of their efficiency programs without large increases to the cost of saved electricity. We discuss examples of energy efficiency forecasting and potential modeling in light our econometric analysis and identify methodological improvements relevant to utilities and grid operators. This paper provides insights into the economics of customer-funded efficiency programs that will support regulators, utilities, and policymakers to utilize energy efficiency as a resource.

**Keywords:** energy efficiency; demand-side management; energy economics; energy policy

## 1. Introduction

Utility customer-funded electricity efficiency programs provide a low-cost electricity resource in 42 states, with investments in 2018 totaling more than USD 5.8 billion and savings exceeding 34 terawatt hours (TWh) [1]. Understanding the cost drivers of customer-funded energy efficiency portfolios can help facilitate regulatory oversight of these programs, help utilities better utilize energy efficiency as a resource, and inform building energy decarbonization policies.

Insight into differences between the cost of saved electricity across states and utilities can help regulators benchmark performance and set appropriate spending levels for utilities in their jurisdictions. Similarly, utilities can use this information to develop least-cost investments in efficiency. In particular, regulators and utilities can benefit from understanding how the cost of saved electricity relates to the scale of the efficiency resource portfolios. For example, increased savings levels may introduce operational efficiencies that reduce the cost of saved electricity. However, historical investments in certain efficiency measures may reduce the availability of low-cost efficiency savings and could counteract this effect. Quantification of these economies and diseconomies of scale, therefore, would be valuable to regulators and utilities estimating the remaining efficiency potential and forecasting spending.

Policymakers can leverage an understanding of drivers of the cost of saved electricity to inform long-term decarbonization strategies. First, insight into the cost of saved electricity can help grid operators assess the role of energy efficiency as a demand-side resource alongside traditional supply-side generation resources. For example, the New England Independent System Operator (ISO New England) uses the cost of saved electricity reported by efficiency program administrators (the organizations that implement energy efficiency programs funded by electricity and gas customers) as part of its efficiency forecasting methodology [2]. Second, when the cost of saved electricity is coupled with information on the timing of efficiency savings, policymakers can better evaluate efficiency as a capacity resource that can reduce peak demand. For example, energy efficiency measures can provide significant value to the grid during peak demand periods but its role as a peak resource depends on whether it is cost competitive with generation and other distributed energy resources [3].

The economics of customer-funded energy efficiency programs can also inform policies that promote building electrification, demand flexibility, and reduce risks from extreme heat. Electrification efforts may exhibit economies of scale similar to those of efficiency programs, in particular for residential buildings, given that they incentivize similar measures (e.g., heat pumps) and employ similar implementation strategies (e.g., upstream incentives) [4]. Similarly, programs that seek to mitigate the impacts of extreme heat may draw on measures generally associated with energy efficiency, such as building envelope improvements [5]. Recent state-level efforts at managing extreme heat risk have also identified existing efficiency programs as implementation mechanisms [6]. Finally, recent research into the technical potential for demand flexibility in U.S. buildings finds synergies in the co-deployment of efficiency and flexibility measures [7]. Estimates of the economic potential for demand flexibility would, therefore, likely benefit from an understanding of the economics of large-scale efficiency program investments.

Several studies have documented the cost of saved electricity from energy efficiency programs. For example, research has characterized how the cost of saved electricity in programs administered by American investor-owned utilities varies geographically, temporally, and by market sector [8,9]. Similar work has described the cost of saved electricity for publicly owned American utilities [10]. Internationally, researchers have examined the cost of saved electricity for both state and utility-administered energy efficiency programs in Korea [11,12].

Additionally, the literature also includes studies that employ econometric methods to explain variation in the observed cost of saved electricity. Arimura et al., 2012 estimates the effect of past and current energy efficiency program spending based on changes in electricity demand using a parametric function and non-linear model as well as utility-fixed effects [13]. Arimura et al., 2012 then predicts the lifetime savings of efficiency investments made in the study period 1992–2006 and calculates an average cost of saved electricity under a range of discount rates [13].

Hoffman et al., 2017 considers both a linear and quadratic relationship between the program administrator levelized cost of saved electricity and program year using program administrator fixed effects for programs implemented between 2009 and 2013 [14]. Overall, the authors find mixed results on the significance of program year on the cost of saved electricity. The study does not find statistically significant results at a 95% confidence interval when the model only includes the linear program year term. With the addition of a quadratic program year variable, the quadratic term is significant at a 95% confidence interval and the linear term is just barely not significant ($p$-value of 0.059). The negative linear term and positive quadratic term suggests that the cost of saved electricity declines before reaching a minimum and increasing. Hoffman et al. (2017) also considers models of the same structure on the cost of saved electricity for different types of programs (e.g., residential lighting) [14]. Using data reported in Energy Information Administration (EIA) Form 861 for the period 2010–2019, Knight et al., 2022 explores economies of scale in energy efficiency programs by regressing the program administrator levelized cost of saved electricity on each program administrator's first year savings as a percent of its retail sales in a given year [15]. The authors find a downward linear trend, suggesting that energy efficiency programs have economies of scale. Their model, however, only explains 12% of the observed variation in the levelized costs and does not include any fixed effects.

This study continues a line of research that uses econometric techniques to estimate the costs of saved electricity in customer-funded energy efficiency programs. We build on these analytic efforts by considering economies of scale with annual savings as a percent of sales among a broader set of covariates, including construction labor costs and both time and program administrator fixed effects. As Table 1 summarizes, no model in the literature includes all of these covariates. We address a gap in the literature of econometric analyses of the cost of saved electricity by introducing a variable for historical efficiency achievement–cumulative efficiency savings as a percent of sales. We include this variable to explore potential diseconomies of scale that result from the increasing costs of achieving remaining energy efficiency potential. Past research has conceptualized energy investments as a supply curve in which different technologies and approaches deliver different levels of savings at different costs [16]. If historical energy efficiency efforts are focused on the less costly sections of the efficiency supply curve, the remaining potential would be weighted towards the more expensive end of the supply curve. In this case, we would expect increases in cumulative savings as a percent of sales to increase the cost of saved electricity. We incorporate this variable into our regression model and explore how it varies across states and in between customer segments. We employ a weighted least squares model to account for differences in the size of the efficiency programs.

**Table 1.** Comparison of econometric models of the cost of saved electricity.

| | | | | Covariates | | | | |
|---|---|---|---|---|---|---|---|---|
| **Study** | **Data Source** | **Period** | **Level of Analysis** | **PA Fixed Effects** | **Year Fixed Effects** | **Labor Costs** | **Annual Savings as % of Sales** | **Cumulative Savings as % of Sales** |
| Arimura et al., 2012 [13] | EIA Form 861 | 1992–2006 | Portfolio | Yes | Yes | No | No | No |
| Hoffman et al., 2017 [14] | Utility regulator reports | 2009–2013 | Portfolio, lighting programs | Yes | Yes | No | No | No |
| Knight et al., 2022 [15] | EIA Form 861 | 2010–2019 | Portfolio | No | No | No | Yes | No |

PA stands for program administrator.

## 2. Materials and Methods

### 2.1. Energy Efficiency Program Data

We constructed a dataset of program costs, energy savings, and savings lifetimes on gas and electric energy efficiency programs administered by investor-owned utilities, publicly-owned utilities, government agencies, and third parties. We primarily source these data from annual regulatory reports that utilities file with state energy regulators. In some cases, we acquired program data directly from utilities and from databases and dashboards maintained by regulatory agencies.

The format of the reported program data was typically in tables embedded in PDFs, which we transcribed into a collection template. We assigned each program to a standard program type developed by the Lawrence Berkeley National Laboratory [17]. To prepare the data for analysis we implemented a two-staged review process. First, we manually reviewed the transcription of the reported data. Second, we identified potential data-entry errors by flagging any program with unexpectedly low or high first year costs of saved electricity (below USD 0.01/kWh or above USD 50/kWh). We then manually reviewed the data for any flagged program and compared it with its source. Therefore, we did not reject outliers simply because of high or low costs of saved electricity; we based our analysis on what utilities reported and conducted data cleaning procedures to assure that we accurately represented the reported data. We drew on more than 5500 program-years of these data to build a panel dataset of 32 program administrators in 14 states that operated electric efficiency programs between 2010 and 2018. In 2018, these program administrators accounted for 55% of savings reported by investor-owned utilities to the EIA in the same year (EIA 2019).

In our analysis, we used ex ante estimates of gross savings for each program as reported by program administrators to state regulators in annual filings. They did not include free-rider or spillover effects nor did they account for evaluation, measurement, and verification (EM&V) studies that occurred in the reporting year. For costs, we included all of the costs that an energy efficiency program administrator would incur, including costs for administration, marketing, evaluation, and customer incentives. We did not include the costs participants incurred as part of participation nor the cost of any shareholder incentive that may have resulted from program achievement. For each program administrator and year, we calculated the levelized cost of saved electricity, which amortizes the cost of an efficiency program over the lifetime of the program's savings [18]. We define three market sectors: commercial and industrial (C&I) and residential, which also include programmatic offerings for low-income customers, and cross-cutting, which includes programs that serve multiple customer classes.

In Tables 2 and 3, we show the composition of the panel dataset in terms of the states and market sectors. Across the nine years of program performance in our dataset, three states: California, Massachusetts and Pennsylvania account for 55% of spending and nearly 50% of savings. Combined, their 2900 records of program are 52% of all of the programs in the dataset. C&I programs account for about half of savings and spending, with 40% resulting from residential programs and the remainder from cross-cutting programs. The five most common program types in our dataset, in terms of overall gross electricity savings, were residential lighting, codes and standards, general prescriptive C&I, custom C&I, and residential behavioral programs. These program categorizations are consistent with Hoffman et al., 2013 [17]. Together, these programs account for 48% of the reported electricity savings in our dataset.

**Table 2.** Share of aggregate gross electricity savings and program administrator spending by state.

| State | Share of Aggregate Gross Electricity Savings | Share of Aggregate Program Administrator Costs | Program Count |
|---|---|---|---|
| Arizona (AZ) | 4.0% | 2.4% | 131 |
| California (CA) | 24.0% | 26.0% | 1553 |
| Colorado (CO) | 3.8% | 3.1% | 380 |
| Massachusetts (MA) | 12.1% | 19.7% | 696 |
| Maryland (MD) | 5.2% | 6.2% | 400 |
| Michigan (MI) | 9.9% | 7.4% | 340 |
| Minnesota (MN) | 4.9% | 3.5% | 554 |
| North Carolina (NC) | 4.5% | 2.9% | 128 |
| New Jersey (NJ) | 3.9% | 4.8% | 117 |
| New York (NY) | 6.4% | 6.4% | 247 |
| Oregon (OR) | 4.7% | 5.4% | 48 |
| Pennsylvania (PA) | 13.7% | 9.5% | 653 |
| Texas (TX) | 0.6% | 0.6% | 217 |
| Utah (UT) | 2.4% | 2.1% | 86 |

**Table 3.** Share of aggregate gross electricity savings and program administrator spending by market sector.

| Market Sector | Share of Aggregate Gross Electricity Savings | Share of Aggregate Program Administrator Costs | Program Count |
|---|---|---|---|
| Residential | 40.9% | 40.2% | 2156 |
| C&I | 49.4% | 53.0% | 2402 |
| Cross Cutting | 9.7% | 6.9% | 992 |

### 2.2. Annual and Cumulative Savings as a Percent of Sales

We measured program scale relative to a utility's sales base using two metrics: annual savings as a percent of annual retail sales and cumulative savings as a percent of annual retail sales. We calculated annual savings as a percent of sales by dividing annual electric efficiency savings by annual retail sales as reported in EIA 861. Where a program administrator's impacts were broader than one utility jurisdiction, we aggregated the savings and sales from the applicable utilities. To characterize the composition of each program administrator's sales base, we calculated the share of retail sales only from residential and C&I customers in each year. We also measured program scale by customer class by calculating annual residential and C&I savings as a percent of retail sales for each program administrator.

We used cumulative savings as a percent of sales to measure the scale of savings from past investments in energy efficiency that have not reached the end of their useful life. We used reported savings lifetimes to estimate when program savings expire. Program administrators reported lifetimes for half of the programs in our dataset. For the remaining programs, we estimated lifetimes based on the lifetimes reported for the same program type. For a given year, we aggregated the savings from prior years that have not expired, regardless of when they began, and divided them by retail sales in that year. For example, a program administrator's cumulative savings in 2018 would reflect all of the savings from program investments made between 2010 and to 2017 that are still active. If a program administrator reported in 2012 that savings from an efficiency program in that year would last 7 years, we would include those savings in the cumulative savings for each year 2013–2018. Cumulative savings as a percent of sales, therefore, can exceed annual savings as a percent of sales, particularly in later years. We did not include savings from before 2010 in this calculation. While some savings from years prior to 2010 likely continue through our analysis period, we lack the data to quantify their contribution. Program administrator fixed effects in the regression model address this issue as cumulative saving pre-2010 are time invariant characteristics of administrators.

### 2.3. Regression Analysis

#### 2.3.1. Base Model

We quantified drivers of the cost of saved electricity through a regression analysis and built on past econometric efforts by accounting for the impact of cumulative energy efficiency savings. Given that our sample includes both small and large investor-owned utilities (the largest annual electricity sales for a single utility is 26 times the smallest annual sales in our dataset), we employed a weighted least squares model with annual retail sales as weights. We specify the regression model in Equation (1). The subscripts on the covariates indicate the dimensions in which the covariates may vary: time (t) and program administrator (PA). The main covariates of interest are annual savings as a percent of sales (S) and cumulative savings as a percent of sales (C). Furthermore, we used two types of fixed effects in our model. Calendar year fixed effects ($\mu_t$) account for factors that vary in time but are common to each administrator, such as macro-economic conditions or the impact of federal appliance efficiency standards. Program administrator fixed effects ($\eta_{PA}$) account for factors that are static in time but vary with each administrator such as differences in program administration, service territory, or regulatory environment. We also addressed how state-level differences and changes in labor costs may affect the cost of saved electricity with an annual Employment Cost Index ($\gamma_t$) for construction workers published by the Bureau of Labor Statistics [19]. We also included the savings-weighted lifetime ($\lambda_{PA,t}$) of each program administrators' efficiency portfolio in each year of the study period as it is an input into the formula for the levelized cost of saved electricity [18]. Finally, we included an intercept ($\alpha$) to capture baseline costs of energy efficiency program administration.

$$\text{Cost of saved electricity} = \alpha + \mu_t + \eta_{PA} + \gamma_t + S_{t,PA} + C_{t,PA} + \lambda_{PA,t} \quad (1)$$

### 2.3.2. Additional Covariates

We explored the impact of several covariates in addition to those in the base model. First, we considered whether differences in the customer composition of utility jurisdiction affects the cost of saving electricity. Given that C&I programs are generally less costly than residential programs, a larger C&I customer base could provide a reliable source of low-cost efficiency savings [20]. We tested this hypothesis by including the share of total sales from C&I customers as a covariate. To account for differences in the cost of saved electricity between customer segments, we also included the share of each program administrator's annual savings from C&I, residential, and low-income programs.

### 2.3.3. Alternative Dependent Variable

Next, we tested sector-specific versions of our dependent variable (e.g., the levelized cost of electricity for C&I programs only) and adjusted program scale covariates accordingly. A regression of the levelized cost of electricity only for C&I programs, then, would be on annual and cumulative C&I savings as a percent of sales. Since EIA 861 does not separate residential sales into low income and non-low-income as many program administrators do for their programs, we included low-income programs in the residential levelized cost of saved electricity.

### 2.3.4. Alternative Sample

As robustness check against compositional effects, we apply the base model to a sample that covers a shorter time period 2011–2018 and but includes six more program administrators (38 total) than our base sample and a non-panel dataset that covers 2010–2018 and 117 program administrators.

## 3. Results

### *3.1. Cost of Saved Energy over Time*

Figure 1 shows how the cost of saved electricity changes between 2010 and 2018 for all sectors combined and for each sector (C&I and residential) separately. Each data point represents the cost of saved electricity for a program administrator in a given year, with the boxes displaying the interquartile range in each year. The bottom of the box is the 25th percentile, the horizontal bar is the 50th percentile, and top of the box is the 75th percentile. The whiskers extend to the last data points that are within a distance 1.5 times the interquartile range above the top and below the bottom of the box. The diamonds represent any values beyond the whiskers. Overall, the median cost of saved electricity declines by about USD 0.005/kWh from USD 0.026/kWh to USD 0.021/kWh after increasing in the middle of the time period. C&I programs exhibit a similar decline, beginning at USD 0.022/kWh and ending at USD 0.018/kWh. In contrast, we found a slight increase (USD 0.001/kWh) in the median levelized cost of electricity for residential programs between 2010 and 2018. Residential programs also display more variation than C&I programs, largely due to their inclusion of low-income programs in the former, which are typically more expensive than market-rate residential efficiency programs [20].

### *3.2. Scale of Programs*

In Figure 2, we present the program administrator's levelized cost of saved electricity against annual savings as a percent of sales in each year. Each data point represents a program administrator's cost of saved electricity for a single program year between 2010 and 2018 against its retail sales in that year. Program administrators have a wide range of savings levels and costs of saved electricity that reflect efficiency portfolios at different stages of development. Within our sample, regional differences emerge. The program administrators with the lowest and highest levels of savings are largely in the South and Northeast, respectively. In both regions, however, there is a wide range of costs of saved electricity. Overall, Figure 2 does not reveal a simple relationship between program scale and the cost of saved electricity.

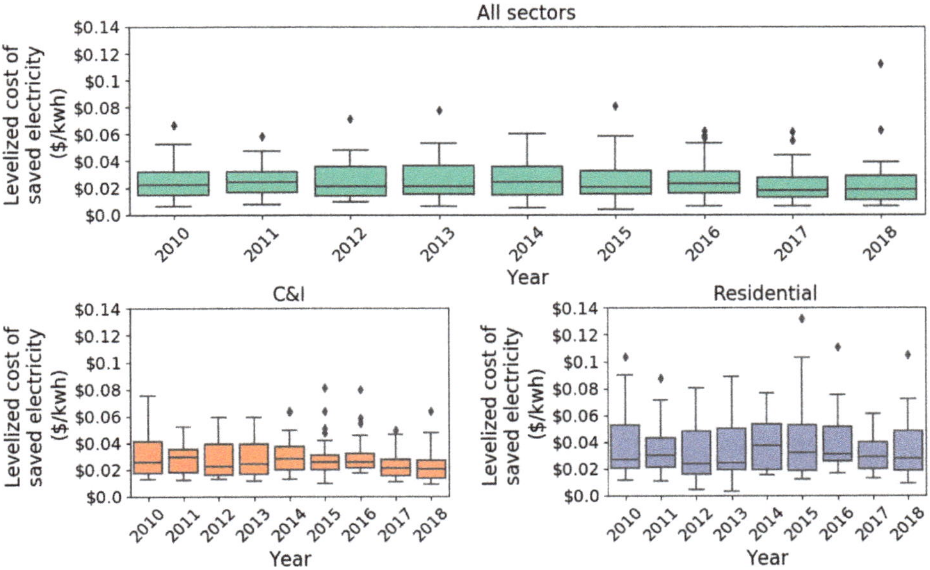


**Figure 1.** Portfolio-wide and sector-level cost of saved electricity 2010–2018.

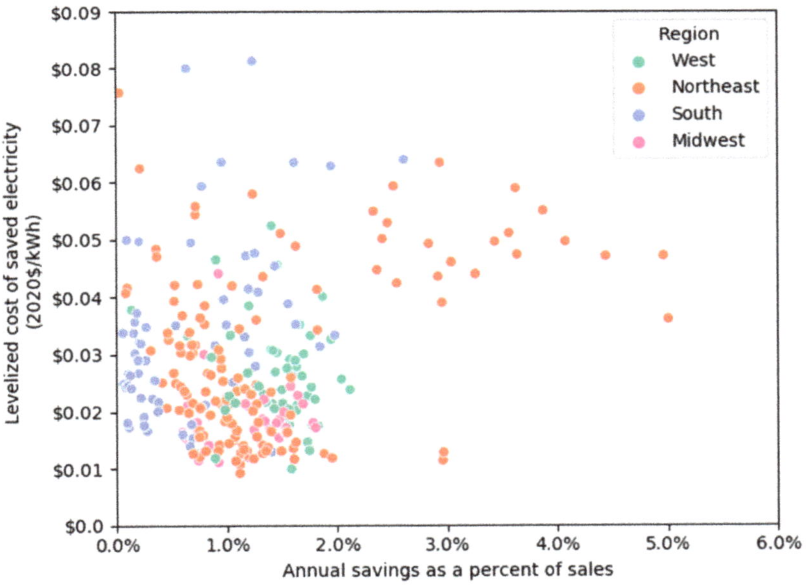


**Figure 2.** Levelized cost of saved electricity by annual savings as a percent of sales 2010–2018.

Figure 3 shows cumulative savings as a percent of sales by state and customer segment in 2018. Differences in annual savings between administrators compound over time, resulting in a wide spread of cumulative efficiency savings by the end of our study period. For example, program administrators in Massachusetts have the highest levels of cumulative savings across all customer segments, equivalent to nearly 20% of retail sales in 2018. In contrast, program administrators in Texas have cumulative savings levels less than two percent of retail sales. In most states in our sample for 2018, cumulative residential savings as a percent of residential sales are higher than cumulative C&I savings as a percent of C&I sales.

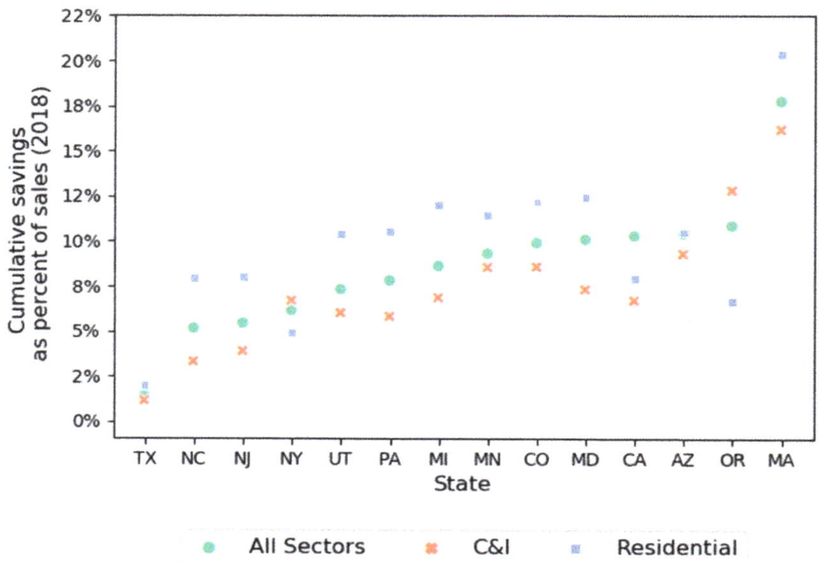


**Figure 3.** Cumulative savings as percent sales by state.

These differences accord with the overall relationship that we observed between sales and savings for residential and C&I programs in our dataset. In 2018, C&I customers account for 65% of sales and 46% of first year savings, whereas residential customers account for 35% of sales and 39% of savings. This trend suggests that significant commercial energy efficiency opportunities may remain for C&I customers in the states in our sample.

### 3.3. Regression Results

We found that energy efficiency programs in our base sample display both economies and diseconomies of scale, are sensitive to construction labor costs, and exhibit both year and program administrator fixed effects. Together, these factors explain 64% of the observed variation in the levelized cost of saved electricity in our dataset. We summarize these results in Table 4.

As the scale of the program (measured as the annual savings as percent of sales) increases, we found that the levelized cost of saved electricity decreases. For a 1% increase in annual savings as a percent of sales, all else being equal, our model estimates a USD 0.006/kWh decrease in the levelized cost of saved electricity. In contrast, the cost of saved electricity increases by about USD 0.14/kWh for a 1% increase in cumulative savings as a percent of sales. The positive return to scale from an increase in annual savings levels, importantly, is more than four times the negative return to scale of an equivalent increase in the cumulative savings level. Since cumulative savings as a percent of sales can increase while annual savings as a percent of sales remains constant, these results suggest that the diseconomies of scale could offset the economies of scale. However, while annual savings as percent of sales is a statistically significant predictor of the levelized cost of saved electricity in the 2011–2018 sample, cumulative savings as a percent of sales is not (see Appendix A for results of alternative model specifications). Additionally, we found that annual savings as a percent of sales is a significant predictor for the residential and C&I levelized cost of saved electricity (see Appendix A for regression results). Cumulative savings as a percent of sales is nearly a significant predictor for the C&I and not for the residential levelized cost of saved electricity at a 95% confidence interval. Overall, our findings provide strong evidence that energy efficiency programs exhibit economies of scale and weak evidence for diseconomies of scale during this time period.

**Table 4.** Program administrator and program year fixed-effects model with weighted least squares.

| | Levelized Cost of Saved Electricity<br>All Sectors |
|---|---|
| Annual savings as percent of sales | −0.591 *** |
| | (0.152) |
| Cumulative savings as percent of sales | 0.137 ** |
| | (0.061) |
| Savings lifetime | −0.001 *** |
| | (0.000) |
| Employment Cost Index | 0.025 *** |
| | (0.006) |
| Intercept | 0.029 *** |
| | (0.003) |
| Observations | 288 |
| $R^2$ | 0.637 |
| Adjusted $R^2$ | 0.575 |
| Residual Std. Error | 40.285 (df = 245) |
| F Statistic | 46.330 *** (df = 42; 245) |

Note:** $p < 0.05$; *** $p < 0.01$.

Next, we found that the 17% growth in the construction Employment Cost Index between 2010 and 2018 increases the levelized cost of saved electricity by about USD 0.005/kWh. For the C&I levelized cost of saved electricity, the index is also a statistically significant predictor and has a somewhat larger effect on costs over the study period (USD 0.008/kWh). We did not find a statistically significant impact of the index for the residential levelized cost of saved electricity. This result may reflect the differences in the labor markets of commercial and residential energy efficiency programs. Commercial programs may draw on a workforce more similar to construction jobs tracked by the Employment Cost Index than do the residential programs.

We also found statistically significant program administrator fixed effects in our base model and across all alternative specifications and samples. Each effect is relative to a base program administrator, National Grid Massachusetts, which we selected because it had the highest cost of saved electricity among program administrators in our sample. To explore whether program administrator effects were only different from just National Grid Massachusetts or more generally from each other, we performed pairwise t tests of each effect's regression coefficient against each other's effect regression coefficient. For each program administrator, we found a significant difference at a 99% confidence interval between its effect and those of the majority of the other program administrator effects. These tests indicate that the program administrator effects are not just different from those of our base program administrator, but from each other more broadly. Furthermore, they underscore the importance of the program administrator effects in explaining variation in the cost of saved electricity. We summarize the program administrator fixed effects results in Appendix A.

The year fixed effects show a decline in the cost of saved electricity across all of the program administrators in our base model, all else equal. This trend of declining fixed costs may reflect the maturation of the energy efficiency industry in our sample through the availability of lower cost vendors and equipment. Full results for the year fixed effects are in Appendix A. We found that these results also hold for the 2011–2018 sample, but not for the non-panel dataset, which may reflect the non-panel dataset's imbalanced coverage of the time period.

Unsurprisingly, we found that the savings lifetime has a small but significant effect on the levelized cost of saved electricity. For a one year increase in savings lifetime, the levelized cost of saved electricity declines by USD 0.001/kWh. We would expect this result by construction since levelized cost is a function of spending, savings, and lifetime.

## 4. Discussion

Our regression analysis generally aligns with the results of similar work and provides new evidence for other drivers of the cost of saved electricity. First, we found a strong economies of scale: as annual savings as a percent of sales increases, the cost of saved electricity decreases at the portfolio level and for residential and C&I programs separately, all else being equal. In contrast, Knight et al., 2021 concludes that annual savings as a percent of sales and the cost of saved electricity only displays a weak relationship [15]. Additionally, we found some evidence for diseconomies of scale in customer-funded efficiency programs through our estimate of cumulative efficiency savings, though the result was not robust to our alternative model specifications and warrants additional research. Together with employment costs and program administrator and year fixed effects, these measures of program scale explain the majority of the observed variation in the cost of saved electricity ($R^2 = 0.637$), much higher than the 12% explained by the model in Knight et al. [15]. Future research could build on this work by exploring the impact of various policy variables on the cost of saved electricity. For example, higher avoided costs would allow more costly programs to pass cost-effectiveness tests, thereby putting upward pressure on the cost of saved electricity. Conversely, higher discount rates in screening tests could select for less costly programs.

Our regression results demonstrate the importance of programs scale-as measured by annual savings as a percent of sales-on the cost of saved electricity. Energy efficiency forecasting and potential study methodologies, however, do not always reflect this relationship. For example, in its energy efficiency forecasts, ISO New England uses a deemed cost escalation factor that reflects measure penetration and technology costs. The escalation factor is graduated, beginning at 0% in the first year of the forecast and increasing by 1.25% each year thereafter [2,21]. These factors linearly increase the cost of saved electricity in each year of the forecast, regardless of any economies of scale. Since ISO New England forecasts savings as a function of budgets and the cost of saved electricity, these escalation factors put downward pressure on savings forecasts. The results of our regression analysis, however, suggest a more nuanced approach to estimating the cost of saved electricity is warranted. In the context of ISO New England's forecasts, annual and cumulative savings as a percent of sales, program administrator effects, and labor costs may all contribute to more accurate cost projections.

Simplistic assumptions about the cost of saved electricity also affect energy efficiency potential studies. As part of its 2020 integrated resource plan, Duke Energy commissioned a study of its energy efficiency savings potential. The study included an 'Enhanced Scenario' in which incentives double but energy savings only increase by 6.3% on average relative to a base case scenario [22]. The economies of scale that we found imply that such significant increases in incentive levels are not necessary for higher levels of savings. Similarly, a 2020 efficiency potential study for Evergy assumes that the first year cost of saved electricity declines as savings levels increase for most commercial programs but not for residential programs [23]. Our results, however, suggest that economies of scale apply to both residential and commercial programs. In general, models of energy efficiency savings potential should account for economies of scale. However, we are less confident in applying the diseconomies of scale from our regressions to efficiency potential models since this result was weaker than that for the economies of scale.

These examples of how energy efficiency forecasts and potential studies are illustrative but not necessarily representative. Still, they illustrate how empirical estimates of the cost of saved electricity are relevant for utilities, regulators, and policymakers. Additional research into energy efficiency forecasting and potential study methodologies would help characterize industry trends in projecting the cost of saved electricity. Programmatic efforts for demand flexibility, building electrification, and extreme heat mitigation may also find these results relevant given potential similarities in implementation to efficiency programs.

The differences we observed in state-level cumulative savings may reflect the existence of an efficiency achievement gap. Importantly, energy efficiency savings do occur outside of customer-funded programs, including through building codes, appliance and equipment standards, and private investment. Higher rates of investments in energy service companies' (ESCO) projects may compensate for lower customer-funded savings levels. For example, New York is in the bottom half of cumulative savings achievement in our sample but has relatively high rates of ESCO investment [24]. In contrast, Texas has the lowest cumulative savings in our sample and only moderate levels of ESCO investment, which suggests that the state may have significant remaining efficiency potential [24]. The economies of scale in our regression model suggest that Texas or a similar state with lower levels of overall energy efficiency achievements can increase efficiency savings goals without significant increases and possible with decreases in the cost of saved electricity.

Our analysis produces quantitative results that are specific to energy efficiency programs administered by investor-owned utilities in the U.S. However, some qualitative results are relevant to policymakers in other countries that have utility- or state-led energy efficiency programs [11,12,25]. In particular, our evidence for economies of scale may be informative to policy conversations on levels of investment and be relevant to jurisdictions developing energy efficiency forecasts or potential estimates. While our dataset includes program administrators that account for more than half of energy efficiency savings reported by investor-owned utilities to the IEA in 2018, it does not include administrators from every state. Additionally, our dataset covers the years 2010–2018 and does not include data from efficiency programs operating in the wake of the COVID-19 pandemic, during which lockdowns shifted electricity usage towards residential buildings [26] and the energy efficiency sector experienced significant job losses [27]. Analysis of an expanded panel dataset that includes data for the years 2019–2022 would shed light on whether the effects we observed for 2010–2018 have continued.

## 5. Conclusions

In this study, we leveraged nearly a decade of data on utility customer-funded energy efficiency programs to examine drivers of the cost of saved electricity through a regression model. We found strong evidence for economies of scale and weaker evidence for diseconomies of scale in program energy savings levels. These results are relevant to system operators forecasting efficiency, utilities estimating efficiency potential, and policymakers considering investments in energy efficiency or exploring strategies for building electrification, demand flexibility, and extreme heat mitigation. Furthermore, we documented large differences in state-level cumulative energy efficiency savings that may suggest an efficiency achievement gap. Our regression analysis suggests that states' savings can overcome this gap and increase energy efficiency savings levels without significantly increasing the cost of saved electricity. Future research can build on this work by including efficiency program data from more states and from more recent years, including those after the COVID-19 pandemic. Such research could test whether the effects we observed for 2010–2018 have continued and could explore the significance of cumulative savings as a percent of sales on the cost of saved electricity.

**Author Contributions:** Conceptualization, S.M.; methodology, S.M.; software, S.M.; data analysis, S.M.; literature review, S.M.; writing—original draft preparation, S.M.; writing—review and editing, S.M. and N.M.F.; project administration, S.M.; funding acquisition, N.M.F. All authors have read and agreed to the published version of the manuscript.

**Funding:** The work described in this study was supported by the U.S. Department of Energy under Lawrence Berkeley National Laboratory Contract No. DE-AC02-05CH11231.

**Data Availability Statement:** The data presented in this study are available on request from the corresponding author.

**Acknowledgments:** The authors thank David Nemtzow from the U.S. Department of Energy for supporting this project.

**Conflicts of Interest:** The authors declare that they have no known competing financial interests or personal relationships that could have appeared to influence the work reported in this paper.

## Appendix A. Regression Results

**Table A1.** Full regression results for base model.

| | Levelized Cost of Saved Electricity All Sectors |
|---|---|
| Annual savings as percent of sales | −0.591 *** |
| | (0.152) |
| Cumulative savings as percent of sales | 0.137 ** |
| | (0.061) |
| Savings lifetime | −0.001 *** |
| | (0.000) |
| Employment Cost Index | 0.025 *** |
| | (0.006) |
| Intercept | 0.029 *** |
| | (0.003) |
| 2010 | 0.017 *** |
| | (0.006) |
| 2011 | 0.019 *** |
| | (0.005) |
| 2012 | 0.014 *** |
| | (0.005) |
| 2013 | 0.013 *** |
| | (0.005) |
| 2014 | 0.013 *** |
| | (0.004) |
| 2015 | 0.008 *** |
| | (0.002) |
| 2016 | 0.007 *** |
| | (0.002) |
| 2017 | 0.002 ** |
| | (0.001) |
| Southern California Edison | −0.021 *** |
| | (0.002) |
| Southwestern Public Service Company-Texas | −0.027 *** |
| | (0.006) |
| West Penn Power | −0.036 *** |
| | (0.004) |
| Western Massachusetts Electric Company | 0.006 *** |
| | (0.000) |
| Consolidated Edison Company of New York | −0.021 *** |
| | (0.004) |
| Consumers Energy Company | −0.025 *** |
| | (0.003) |
| Delmarva Power and Light-Maryland | 0.002 |
| | (0.004) |
| Detroit Edison Company, The | −0.036 *** |
| | (0.003) |
| Duke Energy-North Carolina | −0.035 *** |
| | (0.004) |
| Duquesne Light Company | −0.032 *** |
| | (0.004) |
| El Paso Electric-Texas | −0.026 *** |
| | (0.005) |

**Table A1.** *Cont.*

| | Levelized Cost of Saved Electricity All Sectors |
|---|---|
| Energy Trust of Oregon | −0.022 *** |
| | (0.003) |
| Entergy Texas | −0.024 *** |
| | (0.006) |
| Arizona Public Service Company | −0.036 *** |
| | (0.002) |
| Metropolitan Edison Company | −0.036 *** |
| | (0.003) |
| Minnesota Power | −0.038 *** |
| | (0.004) |
| NSTAR | −0.006 *** |
| | (0.001) |
| Baltimore Gas and Electric Corporation | −0.022 *** |
| | (0.003) |
| New Jersey Clean Energy Program | −0.020 *** |
| | (0.004) |
| New York State Electric & Gas Corporation | −0.017 *** |
| | (0.005) |
| Northern States Power-Minnesota | −0.032 *** |
| | (0.003) |
| PECO Energy | −0.029 *** |
| | (0.004) |
| PPL Electric Utilities Corporation | −0.031 *** |
| | (0.003) |
| Pacific Gas and Electric Company | −0.021 *** |
| | (0.003) |
| Pennsylvania Electric Company | −0.032 *** |
| | (0.003) |
| Pennsylvania Power Company | −0.035 *** |
| | (0.003) |
| Potomac Edison Company-Maryland | −0.009 ** |
| | (0.004) |
| Public Service Company of Colorado | −0.030 *** |
| | (0.003) |
| Central Hudson Gas & Electric Corporation | −0.020 *** |
| | (0.004) |
| Rochester Gas and Electric Corporation | −0.024 *** |
| | (0.004) |
| Rocky Mountain Power-Utah | −0.029 *** |
| | (0.004) |
| Observations | 288 |
| $R^2$ | 0.637 |
| Adjusted $R^2$ | 0.575 |
| Residual Std. Error | 40.285 (df = 245) |
| F Statistic | 46.330 *** (df = 42;245) |

Note: ** $p < 0.05$; *** $p < 0.01$. Program administrator fixed effects are relative to National Grid as base. Year fixed effects are relative to 2018 as base.

**Table A2.** Regression results for C&I programs only.

| | Levelized Cost of Saved Electricity C&I |
|---|---|
| Annual C&I savings as a percent of sales | −1.664 *** |
| | (0.594) |
| Cumulative C&I savings as a percent of sales | 0.130 * |
| | (0.067) |
| Savings lifetime—C&I | −0.002 ** |
| | (0.001) |
| Employment Cost Index | 0.043 *** |
| | (0.013) |
| Intercept | 0.036 *** |
| | (0.007) |
| Observations | 288 |
| $R^2$ | 0.497 |
| Adjusted $R^2$ | 0.411 |
| Residual Std. Error | 88.638 (df = 245) |
| F Statistic | 17.297 *** (df = 42; 245) |

Note: * $p < 0.1$; ** $p < 0.05$; *** $p < 0.01$.

**Table A3.** Regression results for residential programs only.

| | Levelized Cost of Saved Electricity Residential |
|---|---|
| Annual residential savings as a percent of sales | −1.688 *** |
| | (0.653) |
| Cumulative residential savings as a percent of sales | 0.478 |
| | (0.423) |
| Savings lifetime—residential | −0.003 |
| | (0.002) |
| Employment Cost Index | 0.029 |
| | (0.041) |
| Intercept | 0.048 *** |
| | (0.014) |
| Observations | 288 |
| $R^2$ | 0.288 |
| Adjusted $R^2$ | 0.166 |
| Residual Std. Error | 224.645 (df = 245) |
| F Statistic | 15.740 *** (df = 42; 245) |

Note: *** $p < 0.01$.

## References


1. Consortium for Energy Efficiency. *CEE Annual Industry Report*; Consortium for Energy Efficiency: Middleton, MA, USA, 2020.
2. ISO New England. *Final 2017 Energy-Efficiency Forecast for 2021–2016*; ISO New England: Holyoke, MA, USA, 2017.
3. Murphy, S.; Deason, J. Timed to Save: The Added Value of Accounting for Hourly Incidence of Electricity Savings from Residential Space Conditioning Measures. *Energy Effic.* **2021**, *14*, 82. [CrossRef]
4. Deason, J.; Wei, M.: Leventis, G.; Smith, S.; Schwartz, L.C. *Electrification of Buildings and Industry in the United States: Drivers, Barriers, Prospects, and Policy Approaches*; Lawrence Berkeley National Laboatory: Berkeley, CA, USA, 2018; p. 1430688.
5. Baniassadi, A.; Sailor, D.J. Synergies and Trade-Offs between Energy Efficiency and Resiliency to Extreme Heat—A Case Study. *Build. Environ.* **2018**, *132*, 263–272. [CrossRef]
6. California Natural Resources Agency. *Draft Extreme Heat Action Plan*; California Natural Resources Agency: Sacramento, CA, USA, 2022; p. 46.
7. Langevin, J.; Harris, C.B.; Satre-Meloy, A.; Chandra-Putra, H.; Speake, A.; Present, E.; Adhikari, R.; Wilson, E.J.H.; Satchwell, A.J. US Building Energy Efficiency and Flexibility as an Electric Grid Resource. *Joule* **2021**, *5*, 2102–2128. [CrossRef]
8. Goldman, C.A.; Hoffman, I.; Murphy, S.; Frick, N.A.; Leventis, G.; Schwartz, L.C. The Cost of Saving Electricity: A Multi-Program Cost Curve for Programs Funded by U.S. Utility Customers. *Energies* **2020**, *13*, 2369. [CrossRef]

9. Baatz, B.; Gilleo, A. Big Savers: Experiences and Recent History of Program Administrators Achieving High Levels of Electric Savings. *Electr. J.* **2016**, *29*, 9–16. [CrossRef]
10. Schwartz, L.C.; Hoffman, I.M.; Schiller, S.R.; Murphy, S.; Leventis, G. *Cost of Saving Electricity Through Efficiency Programs Funded by Customers of Publicly Owned Utilities: 2012–2017*; Lawrence Berkeley National Laboatory: Berkeley, CA, USA, 2019; p. 1576511.
11. Kang, H.; Kim, Y.; Baek, J. Estimating the Cost of Saving Electricity of Energy Efficiency Programs: A Case Study of South Korea. *Energy Policy* **2022**, *160*, 112672. [CrossRef]
12. Lee, W.-N.; Kim, H.-J.; Park, J.-B.; Roh, J.-H.; Cho, K.-S. An Economic Evaluation of the Energy Efficiency Programs in Korea. *J. Int. Counc. Electr. Eng.* **2014**, *2*, 219–224. [CrossRef]
13. Arimura, T.H.; Li, S.; Newell, R.G.; Palmer, K. Cost-Effectiveness of Electricity Energy Efficiency Programs. *Energy J.* **2012**, *33*, 63–99. [CrossRef]
14. Hoffman, I.M.; Leventis, G.; Goldman, C.A. *Trends in the Program Administrator Cost of Saving Electricity for Utility Customer-Funded Energy Efficiency Programs*; Lawrence Berkeley National Laboatory: Berkeley, CA, USA, 2017; p. 16.
15. Knight, P.; Biewald, B.; Takahashi, K. The Cost of Energy Efficiency Programs: Estimates from Utility-Reported Datasets. *Energy* **2022**, *239*, 122448. [CrossRef]
16. McKinsey & Company. *Reducing U.S. Greenhouse Gas Emissions: How Much at What Cost?* Mckinsey & Company: New York, NY, USA, 2007.
17. Hoffman, I.M.; Billingsley, M.A.; Schiller, S.R.; Goldman, C.; Stuart, E. *Energy Efficiency Program Typology and Data Metrics: Enabling Multi-State Analyses through the Use of Common Terminology*; Lawrence Berkeley National Laboratory: Berkeley, CA, USA, 2013.
18. Meier, A. *The Cost of Conserved Energy as an Investmentt Statistic*; Lawrence Berkeley National Laboratory: Berkeley, CA, USA, 1982; p. 23.
19. Bureau of Labor Statistics. *Employment Cost Index*; Bureau of Labor Statistics: Washinton, DC, USA, 2021.
20. Frick, N.M.; Murphy, S.; Miller, C.; Pigman, M. *Still the One: Efficiency Remains a Cost-Effective Electricity Resource*; Lawrence Berkeley National Laboratory: Berkeley, CA, USA, 2021; p. 43.
21. ISO New England. *Final 2022 Energy Efficiency Forecast*; ISO New Enland: Holyoke, MA, USA, 2022.
22. Nexant. *Duke Energy Carolinas Integrated Resource Plan Attachment V: Duke Energy EE and DSM Market Potential Study*; Duke Energy: Charlote, NC, USA, 2020.
23. ICF. *Evergy 2019 DSM Potential Study*; ICF: Kansas City, MO, USA, 2020.
24. Larsen, P.; Carvallo, J.P.; Murphy, S.; Stuart, E.; Goldman, C. *State of the U.S. ESCO Industry—Market Characteristics and Project Performance: 1990–2017*; Lawrence Berkeley National Laboratory: Berkeley, CA, USA, 2019.
25. Abhyankar, N.; Phadke, A. Impact of Large-Scale Energy Efficiency Programs on Utility Finances and Consumer Tariffs in India. *Energy Policy* **2012**, *43*, 308–326. [CrossRef]
26. Krarti, M.; Aldubyan, M. Review Analysis of COVID-19 Impact on Electricity Demand for Residential Buildings. *Renew. Sustain. Energy Rev.* **2021**, *143*, 110888. [CrossRef] [PubMed]
27. Jordan, P. *Clean Energy Employment Initial Impacts from the COVID-19 Economic Crisis*; BW Research Partnership: Carlsbad, CA, USA, 2020.

*energies*

MDPI

*Article*

# Enhanced Power Quality in Single-Phase Grid-Connected Photovoltaic Systems: An Experimental Study

**Abdellah Benabdelkader [1], Azeddine Draou [2], Abdulrahman AlKassem [2], Toufik Toumi [1], Mouloud Denai [3], Othmane Abdelkhalek [1] and Marwa Ben Slimene [4,5,*]**

1 Smart Grids & Renewable Energies Laboratory, Faculty of Technology, University of Tahri Mohammed Bechar, Bechar 08000, Algeria
2 Department of Electrical Engineering, College of Engineering, Islamic University of Madinah, Madinah 42351, Saudi Arabia
3 School of Physics, Engineering & Computer Science, University of Hertfordshire, Hatfield AL10 9AB, UK
4 Department of Computer Science and Engineering, College of Computer Science and Engineering, University of Ha'il, Ha'il 55436, Saudi Arabia
5 SIME Laboratory, ENSIT, University of Tunis, 05 Ave Taha Hussein, Tunis 1008, Tunisia
* Correspondence: benslimene.marwa@gmail.com

**Abstract:** The main aim of the research work presented in this paper consists of proposing an effective control scheme for a grid-connected single-phase photovoltaic (PV) system to enhance not only the power quality at the point of common coupling (PCC) but also to operate with a maximum power point tracking (MPPT) controller. Moreover, an orthogonal signal generator (OSG) module for effective grid synchronization, a current reference generation controller, and a PWM generating block have also been designed and included in this paper. The proposed control strategy allows the MPPT controller to switch to faulty mode and maintains the voltage according to network requirements using an adaptive neuro-fuzzy inference system (ANFIS)-based control whenever a fault occurs at the PCC. The performance of the analyzed control strategy, which is based on the static compensation of the DC-link voltage fluctuations in a grid-connected inverter powered by PV, is further explored through simulations in MATLAB, and the results are included in this paper. Moreover, the control scheme is implemented experimentally using a dSPACE DS 1104 control board and then assessed on a small laboratory-scale single-phase PV system that is subjected to some fault scenarios. The simulation and experimental results have shown improved power quality and robustness against grid fluctuations, resulting in better dynamic performance.

**Keywords:** photovoltaic; orthogonal signal generator; ANFIS; MPPT; DC-AC power quality

## 1. Introduction

In recent years, power quality (PQ) has become a real problem for both utility providers and consumers due to the widespread interconnection of power-electronic-based nonlinear loads to the electricity distribution grid. Voltage sags and swells are among the most common PQ issues affecting consumers' critical loads [1–4]. A voltage sag is defined as a drop in the AC voltage's RMS value at a power frequency of 0.5 cycles per minute, according to IEEE 1346 [5] and IEEE 1159 [5–7] standards. It can also be defined as a rapid drop and simultaneous fast recovery in the voltage at any point of the electrical system according to the standard IEC 61000-2-1 [8]. Voltage sags are often caused by short-circuit faults such as single line-to-ground faults due to the sudden interconnection of a large load such as an industrial induction motor. A voltage swell is described by the IEEE standards as a rise in the RMS value of the supply voltage from 1.1 p.u. to 1.8 p.u. of the nominal voltage during a period from 0.5 s to one minute. The interconnection of large capacitors or the disconnection of large loads are the two primary sources of voltage surges. As a result, some of the industrial equipment, such as motor drives, surge arresters, and relays, may be overheated, tripped, or even damaged by voltage surges [8,9], whereas, for sensitive loads such as electronic equipment, harmonic pollution is of major concern because its regulation is based on the crossings of the zero or maximum value point of the supply voltage [3–7].

Renewable energy generation technology based on power electronics is developing rapidly. New energy generation technology is based on photovoltaic (PV) systems, which are generally located in mountainous areas and deserts to receive more sunlight. In addition, these remote areas are conducive to the large-scale installation of PV inverters [10,11]. Recently, an increase in the deployment of solar PV systems, from rooftop installations to large-scale solar farms in the megawatt range, has been driven not only by the continuous fall in the price of PV modules but also by the consumers' rising awareness of and wish to adopt eco-friendly energy solutions [9]. Future PV systems are expected to meet power generation needs from a point closer to the point of consumption [12–14] to avoid transmission losses and enhance the efficiency of the distribution network. Although there is an extensive body of literature on PV systems, this area is still an active area of ongoing research and development. Solar energy plays a leading role in reducing greenhouse gas emissions and has received strong support through various incentive programs, although the price is still expensive.

In previous research [15,16], the authors describe a modeling technique and fuzzy logic-based MPPT controller for efficient maximum power extraction. The authors of [17] presented an accurate method to adjust the duty cycle under different climatic conditions. The switching pulses of the power converter semiconductor devices, which regulate the active and reactive forms of power, are generated using a peak current controller [18–20]. Future PV systems should be capable of managing reactive power, islanding detection, power quality enhancement, and fault ride-through (FRT), as described in earlier studies [21–23]. In grid-connected inverter systems, some of the voltage amplitudes and frequencies are fundamentally critical parameters; accurate knowledge of these parameters is vital for obtaining the appropriate reference signals and meeting the requirements of future grids [24,25].

In another study [26], the authors have shown that power electronic equipment consumes large quantities of reactive power and can lead to the introduction of harmonics into the network. The authors of [27] have proposed a conventional PI controller to enhance the power quality of a multilayer single-phase inverter. The authors of [28,29] have stressed that voltage sag is a frequent problem that often results in unnecessary tripping when applied to a single-phase system operating in both ON and OFF grid modes. In addition, in an earlier study [30], the authors use a PV system to solve a problem related to the efficient control of the reactive power caused by using loads that influence the current supplied from the source. In islanded mode, the voltage controller maintains the sinusoidal waveforms of the load voltage. In various other studies [31–33], the authors stated that the synchronization controller ensures the transition between the bidirectional islanded and grid-connected modes, depending on the grid's accessibility.

The work presented in this paper intends to carry out experimental analysis on a laboratory prototype to assess the dynamic performance of single-phase grid-connected PV systems that are subjected to various faults. One of the merits of this research is to implement a PV system controller to regulate the active and reactive powers required by the load. A perturb and observe (P&O) and maximum power point tracking (MPPT) algorithm is introduced to allow the PV system to operate at its maximum power. In addition, when a fault occurs at the point of common coupling (PCC) under any voltage sag circumstances, the controller will inject a reactive current, and the system may operate in islanded mode.

The rest of the paper is organized as follows. Section 2 gives a description of the proposed system considered in this work. Section 3 describes the power control concepts and Section 4 presents the proposed ANFIS-based power controller. Section 5 is devoted to the various grid synchronization techniques. In Section 6, a series of simulation results are inserted, and in Section 7, the hardware test bed of the grid-connected PV system developed in this paper is described, together with the introduction of a series of experimental test results. Finally, Section 8 summarizes the main conclusions of the paper.

## 2. Description of the System

For PV systems to operate effectively, it is necessary to control various parameters, such as the PV side for maximum power extraction, the inverter side for injecting high-quality current, and the grid side for any additional tasks. The ambient temperature and solar irradiance must be considered during a PV system's design and planning phases. These factors greatly influence the PV system's output power [34,35]. Figure 1 shows the schematic diagram of a system comprising a dual stage with a PV system connected to the PCC through a DC-DC boost converter, grid inverter, and control system as proposed in the literature [7,22]. The system may operate in both grid-connected and islanded modes.

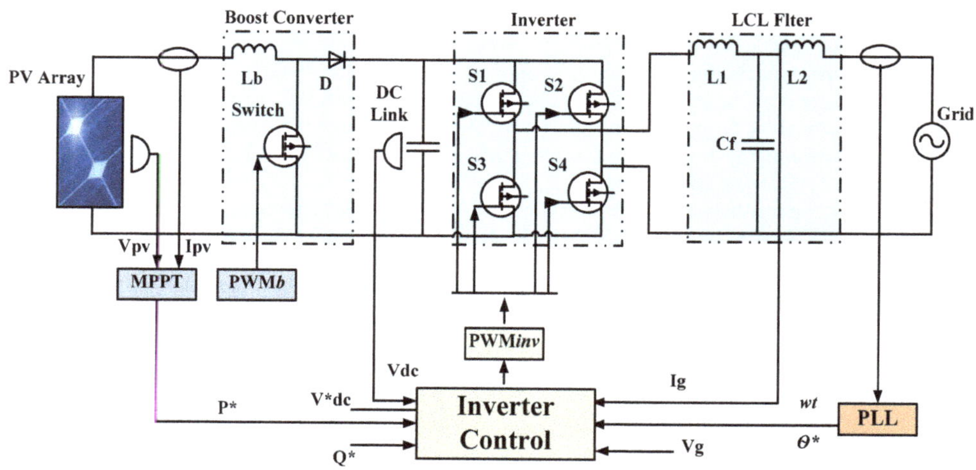


**Figure 1.** Schematic diagram of the grid-connected PV system.

## 3. PV Array Modeling

The PV cell is the basic unit of a PV module; it transforms the photons of the sun's rays directly into electric power. Figure 2 shows the equivalent circuit of a PV system, which consists of N strings of PV modules in parallel with a line consisting of series M modules to achieve the appropriate power rating.

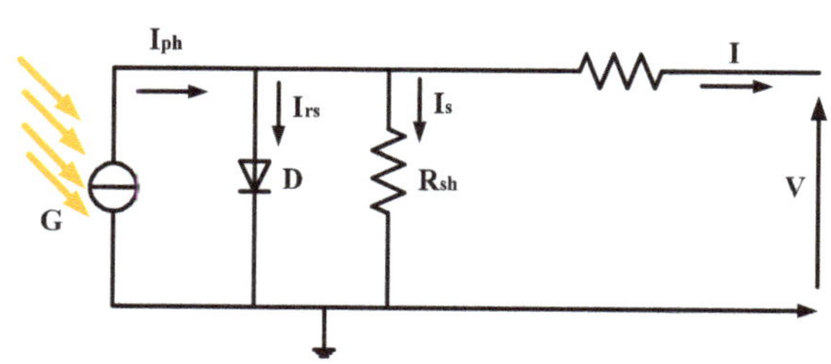


**Figure 2.** Equivalent circuit of a PV cell.

The dynamic model of the PV array is described by the following equations [36].

$$I = N_p I_{ph} - N_p I_0 \left[ \exp\left( \frac{q(V_{pv} + I_{pv} R_{sh})}{N_s AKT} \right) - 1 \right] - \frac{V + IR_{sh}}{R_{sh}} \quad (1)$$

$$I_s = I_{rs} \left( \frac{T_c}{T_{ref}} \right)^3 \left[ \exp\left( \frac{qE_g \left( \frac{1}{T_{ref}} - \frac{1}{T_c} \right)}{KA} \right) \right] \quad (2)$$

$$I_{rs} = \frac{I_s}{\left[\exp\left(\frac{qV_{oc}}{N_s KAT}\right) - 1\right]} \tag{3}$$

$$I_{ph} = [K_i(T - T_{ref})]\left(\frac{G}{1000}\right) \tag{4}$$

To be able to produce maximum power, the PV module must work within a peak value $P_{max}$ of the output power. The MPPT algorithm used in this study is based on the well-known P&O technique, and a 1 kW PV system is considered.

## 4. Analysis of Power Control

This section discusses the control scheme of a single-stage single-phase grid-connected PV system. In normal cases, the PV solar system would be operated in MPPT mode, which may give maximum power to the grid. However, when a fault occurs at the PCC, the grid will require the PV inverter to supply a reactive current to boost it [7]. During islanding, this mode will switch to fault mode, as illustrated by the flowchart diagram in Figure 3.

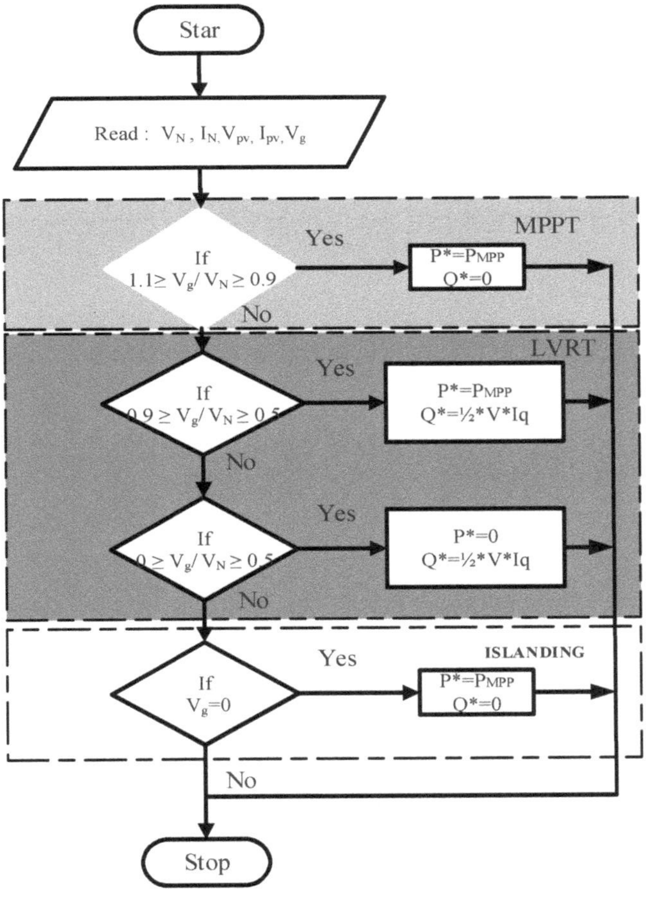


**Figure 3.** Flowchart describing the generation of the reference powers.

The control block diagram of the PV system is shown in Figure 4. It consists of the MPPT, the orthogonal signal generator (OSG) unit, the PWM block, and the PI controllers for current reference generation. A low pass (LP) filter is introduced to smooth the inverter output voltage waveform. In the $EF$ frame, the grid current and voltage components are represented by $I_{gE}$, $I_{gF}$, and $V_{gE}$, $V_{gF}$, respectively. The real ($P$) and reactive ($Q$) powers are defined by the following equations [23] and their values may be calculated using OSG systems in the same manner as in the dq-rotating reference frame:

$$P = \frac{1}{2}(V_{g\alpha}\ I_{g\alpha} + I_{g\beta}\ V_{g\beta}) \quad (5)$$

$$Q = \frac{1}{2}(V_{g\alpha}\ I_{g\alpha} - I_{g\beta}\ V_{g\beta}) \quad (6)$$

in which the currents in the $\alpha\beta$-stationary reference frame are given by:

$$\begin{bmatrix} i_{g\alpha} \\ i_{g\beta} \end{bmatrix} = \frac{2}{v_{g\beta}^2 + v_{g\beta}^2} \begin{bmatrix} v_{g\alpha} & v_{g\beta} \\ v_{g\beta} & -v_{g\alpha} \end{bmatrix} \begin{bmatrix} P \\ Q \end{bmatrix}. \quad (7)$$

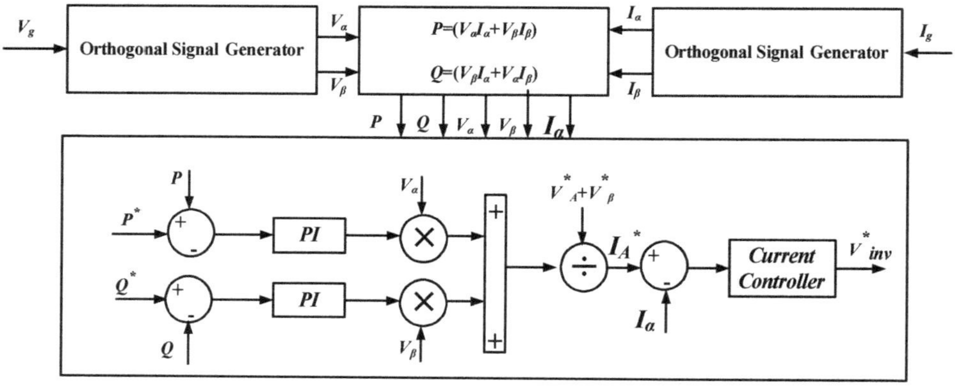


**Figure 4.** Proposed control block diagram of the system based on PQ theory and the OSG concept.

A controller such as a PI may be used to regulate the power since both powers are constant current quantities. The current reference is given by:

$$\begin{bmatrix} i_{g\alpha}^* \\ i_{g\beta}^* \end{bmatrix} = \frac{2}{v_{g\beta}^2 + v_{g\beta}^2} \begin{bmatrix} v_{g\alpha} & v_{g\beta} \\ v_{g\beta} & -v_{g\alpha} \end{bmatrix} \begin{bmatrix} G_P(s)\ (P - P^*) \\ G_q(s)\ (Q - Q^*) \end{bmatrix}. \quad (8)$$

The transfer functions of the PI controllers for both these powers may be expressed as:

$$\begin{aligned} G_P(s) &= K_{pp} + K_{pi}\tfrac{1}{s} \\ G_q(s) &= K_{qp} + K_{qi}\tfrac{1}{s} \end{aligned}. \quad (9)$$

## 5. ANFIS-Based Controller Development

ANFIS belongs to the class of neuro-fuzzy systems, which combine the learning capability and the parallel computation of neural networks with fuzzy systems. Neuro-fuzzy systems can be trained to develop fuzzy rules and determine the membership functions for the variables of the system. The structure of a neuro-fuzzy system is similar to that of a multi-layer neural network and includes an input and an output layer, along with three hidden layers that represent the membership functions and fuzzy rules. ANFIS training uses a hybrid learning algorithm combining the least-squares estimator (forward pass) and the gradient descent method (backward pass) [37]. These features provide ANFIS with superior performance and robustness [38]. In this section, an ANFIS-based controller is proposed for the control of the inverter DC-link voltage. Details of the design steps and a list of the parameters of the ANFIS training are given in Appendix A. Figure 5 depicts the triangular membership functions of the error, $\varepsilon$, and error change, $\Delta\varepsilon$. Figure 6 illustrates the variations of the output against the two inputs $\varepsilon$ and $\Delta\varepsilon$. ANFIS was originally designed with bell-shaped MFs.

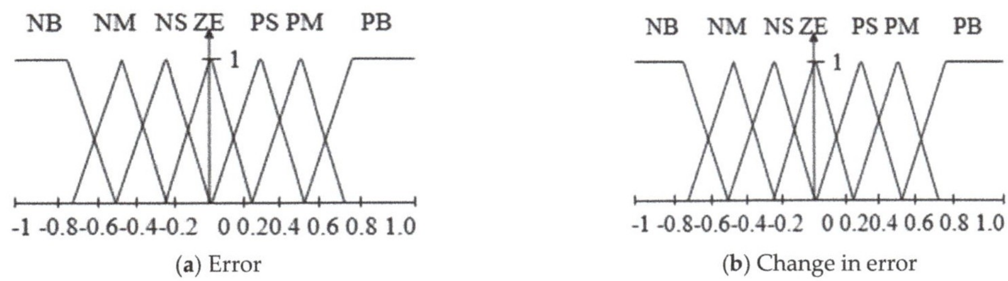


**Figure 5.** Triangular membership functions.

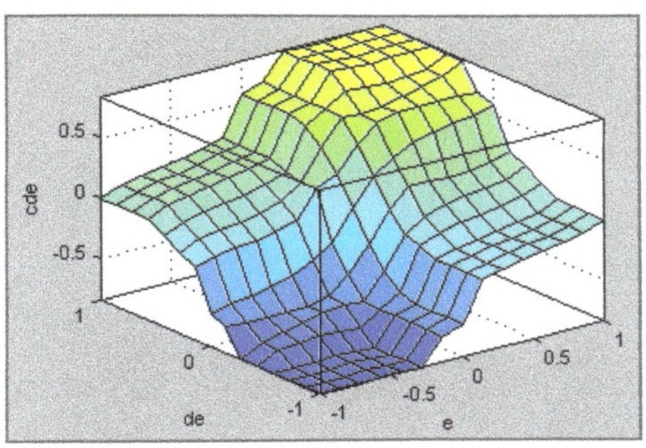


**Figure 6.** Surface view of the ANFIS controller.

## 6. Grid Synchronization Techniques

Following grid failures due to the occurrence of a fault, the synchronization of single-phase PV systems is critical. A system with good synchronization capabilities would have to be designed to respond quickly to any voltage drop. In the recent literature, several synchronization methods have been proposed [38], which may be broadly categorized into two groups:

1. Mathematical methods, e.g., synchronized Fourier- and PLL-based methods. Currently, PLL-based synchronization techniques are the most popular choices since the design of a phase detector, a simple sinusoidal multiplier [38,39]. In a single-phase system, however, this procedure will result in a double-frequency term. Another approach to removing phase error for a PLL-based method is to apply the Park transform to an OSG system. A PLL based on T/4 delay [37] and inverse Park transform-based PLL (IPT-PLL) [40] are examples of such PLLs.
2. Another approach is to employ adaptive filters that can automatically modify the output. The OSG is utilized to help in the calculations of active and reactive powers and provides appropriate grid synchronization. The OSG system is used to calculate $P$ and $Q$. Figure 7 depicts the OSG for a single-phase system [7].

The reactive power is subjected to:

$$I_p^2 + I_q^2 = I_N^2 \leq I_{max}. \tag{10}$$

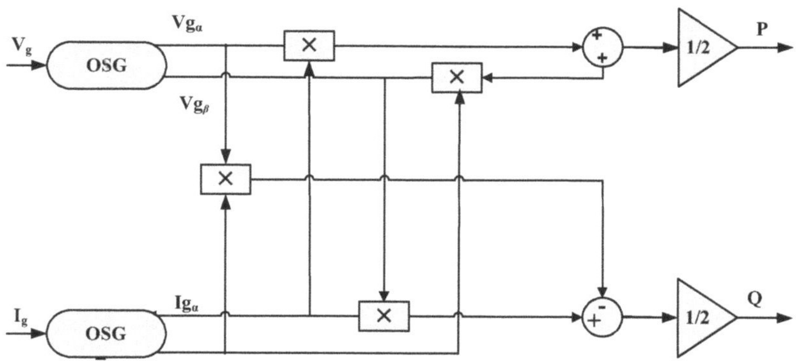


**Figure 7.** Orthogonal signal generation procedure.

The minimum value of the reactive current ($I_q$) to be injected into the grid system is given by Equation (11). The active power is controlled to keep the frequency constant.

$$I_q = \begin{cases} & 0.9 \le \frac{V_g}{V_N} \le 1 \\ 2 \times \frac{V_g}{V_N} \times I_N, & 0.5 \le \frac{V_g}{V_N} \le 0.9 \\ I_N, & 0.9 \le \frac{V_g}{V_N} \le 1 \end{cases} \tag{11}$$

$$I_p = \begin{cases} Im, & 0.9 \le \frac{V_g}{V_N} \le 1 \\ \sqrt{I_N - \left(2 \times \frac{V_g}{V_N} \times I_N\right)}, & 0.5 \le \frac{V_g}{V_N} \le 0.9 \\ 0, & 0.1 \le \frac{V_g}{V_N} \le 0.5 \end{cases} \tag{12}$$

Several recent papers have dealt with control methods to suppress the injected current harmonics. In this case, the harmonic compensation method is used, as shown in Figure 8.

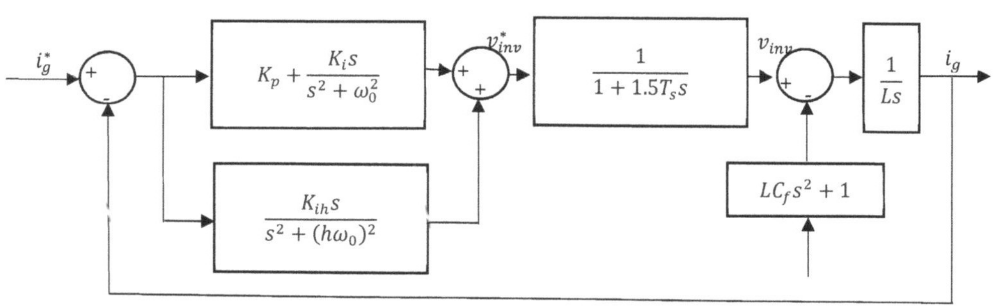


**Figure 8.** Block diagram of the current control loop for harmonic voltage distortion elimination.

Thus, the closed-loop transfer function is obtained as:

$$\frac{i_g(s)}{i_g^*(s)} = \frac{G_{PR+HC}(s)\ G_d(s)}{LS + G_{PR+HC}(s)\ G_d(s)} \tag{13}$$

where $G_{PR+HC}(s)$ represents the controller's transfer function and $G_d(s)$ is the processing time delay.

The relationship of $G_{PR+HC}(s)$ and $G_d(s)$ can also be written as:

$$\begin{gathered} G_{PR+HC}(s) = K_p + \frac{K_i s}{s^2+\omega_0^2} + \sum_{h=3,5,3,7\ldots} \frac{K_{ih} s}{s^2+(h\omega_0)^2} \\ G_d(s) = \frac{1}{1+1.5T_s s} \end{gathered} \tag{14}$$

$$\left( G_{HC}(s) = \frac{K_{ih} s}{s^2 + (h\omega_0)^2} \right) \tag{15}$$

where $k_p$ is the value of the proportional gain and $k_{ih}$ represents the resonant and harmonic compensator gains, respectively; $h$ represents the value of the harmonic order, $\omega_0$ denotes the value of the fundamental frequency, and $T_s$ represents the value of the sampling period.

The current control strategy will decide on the quality of the power injected into the grid and the precise harmonics compensators [7] that are required to achieve this. In the *BC* reference frame, the gains of the PR and HC controllers are defined by Equations (14) and (15) respectively.

The PV system parameters are listed in Table 1. Details of the Simulink implementation of a single-phase grid-connected PV system with an LCL filter is given in Appendix C.

**Table 1.** PV panel, Boost converter, and load specifications.

| Component | Value |
|---|---|
| PV panel maximum power | |
| LC filter values | 2.3 mH, 100 KF |
| Linear load | 450 W, 125 Var |
| Nonlinear load | 850 W, 250 Var |
| Converter values | 2.9 mH, 70 KF, 15 kHz |

## 7. Simulation Results

The performance of the proposed control strategy, applied to a grid-connected inverter powered by PV solar energy, has been analyzed through a simulation created using MATLAB 2016 b. The combined PV array and DC-DC boost converter is connected to the grid through a voltage source inverter. A filter is introduced to obtain a near-sinusoidal voltage waveform.

Figure 9a,b shows the various forms of power during islanding and sag voltage. Moreover, the DC link voltage and real and reactive powers' waveforms for a linear load are depicted in Figure 10a,b, respectively. These results show the good performance of the proposed control system.

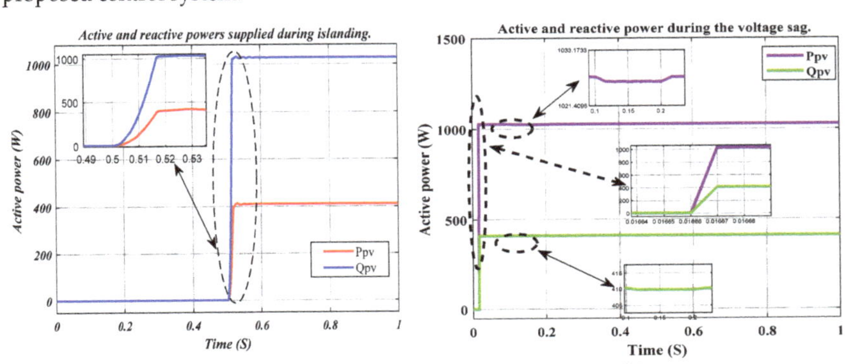


(a) (b)

**Figure 9.** Real and reactive power: (**a**) during islanding, (**b**) during sag voltage.

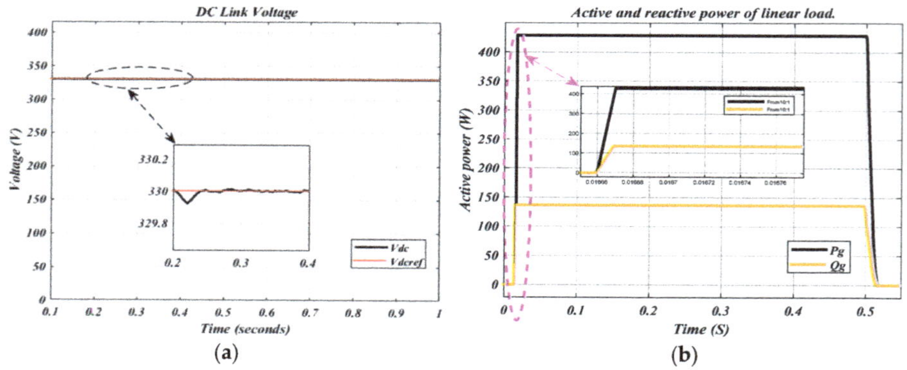


**Figure 10.** Waveforms for a linear load of (**a**) DC link voltage, and (**b**) real and reactive power.

Figure 11a,b shows the various forms of power in the case of power supplied by the source and PV system.

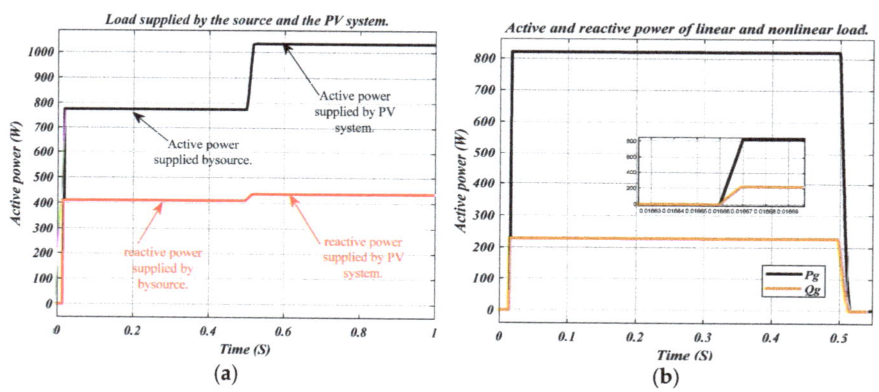


**Figure 11.** Simulation results: (**a**) load supplied by the source and PV; (**b**) real and reactive power for linear and nonlinear loads.

The proposed system is assessed under fault conditions, in the presence of a sag applied at $t = 0.1$ and cleared at $t = 0.2$ s. The results are shown in Figure 12.

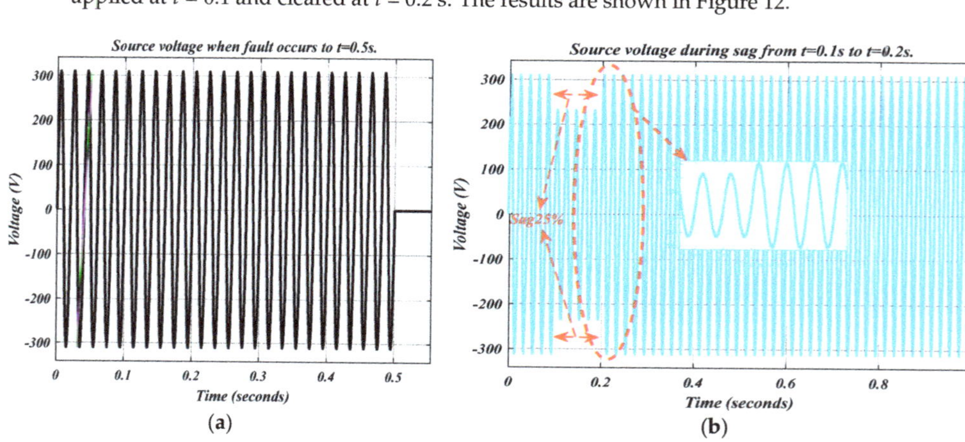


**Figure 12.** Simulation results: (**a**) source voltage under faults; (**b**) source voltage during sag.

Figure 13 present the total harmonic distortion (THD), the grid current with active filtering has been reduced to 2.15%, which can be considered a significant enhancement.

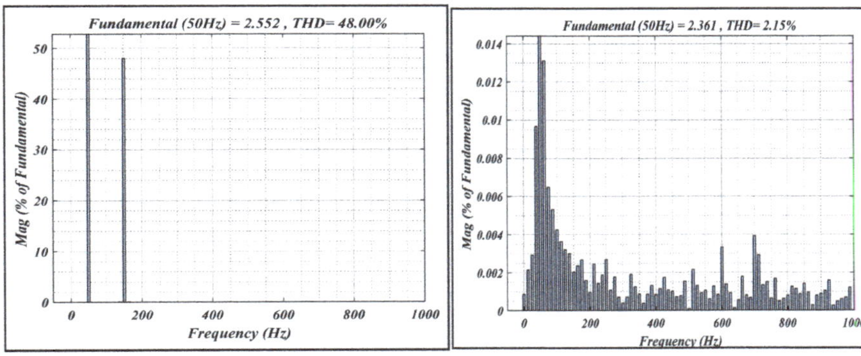


(a) (b)

**Figure 13.** Grid current THD analysis. (**a**) THD before active filtering; (**b**) THD after active filtering.

## 8. Experimental Results

The experimental setup designed to validate the proposed control strategy for the grid-connected PV system is shown in Figure 14. It consists of:

1. SEMIKRON IGBT based inverter;
2. Current and voltage sensors;
3. DC-DC boost converter;
4. Linear load;
5. dSPACE DS 1104 and power quality analyzer;
6. Host PC and oscilloscope;
7. Five PV panels.

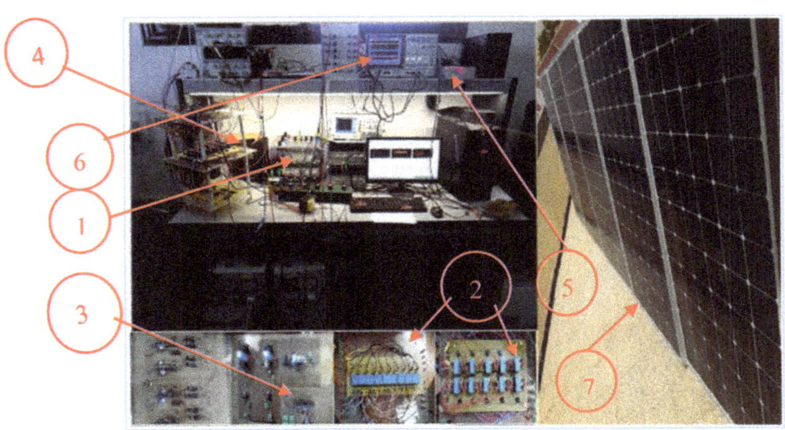


**Figure 14.** Experimental setup for the PV system.

Figure 15 shows the PV panel used in this work, with no load and in operating temperature conditions ranging from 25 °C to 35 °C; the solar irradiance is between 970 W/m$^2$ and 800 W/m$^2$.

(**a**)

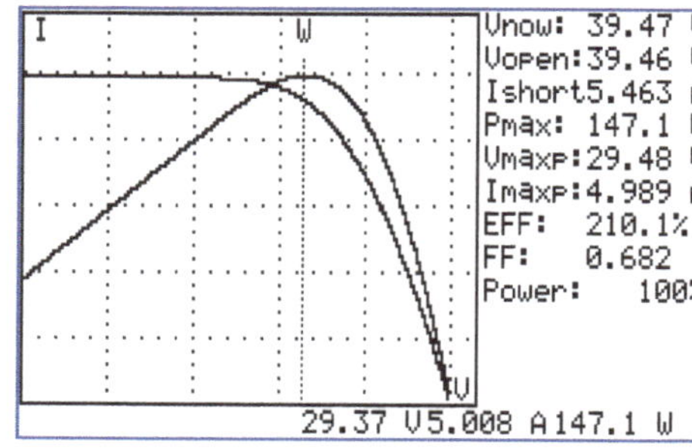


(**b**)

**Figure 15.** Experimental tests: (**a**) irradiance conditions for the solar module; (**b**) solar panel I-V and P-V characteristics.

*8.1. Islanding Mode*

Initially, the boost converter and MPPT controller were designed and tested with linear and non-linear loads for an input voltage of around 200 V and an output voltage of 400 V. Figure 16a shows the experimental results of the PV voltage (185 V) and output voltage (330 V) of the boost converter. As shown in Figure 16b, the duty cycle has a value of 0.7.

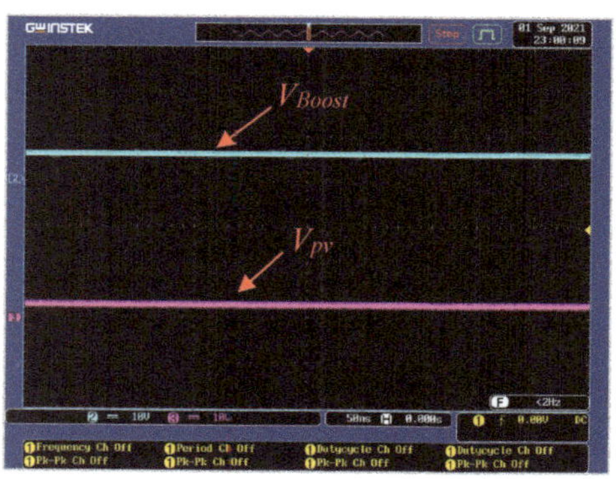


(**a**)

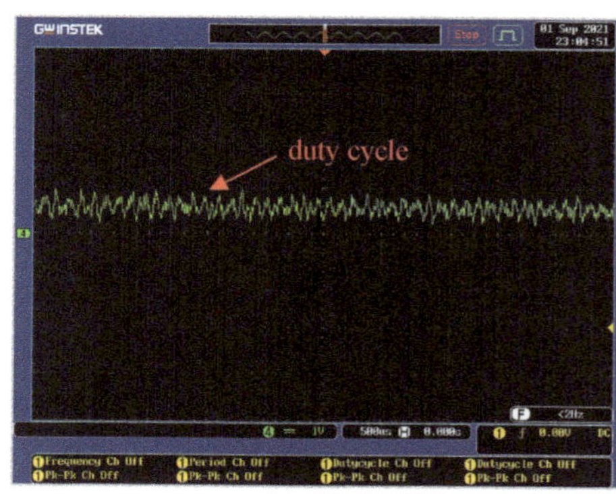


(**b**)

**Figure 16.** Experimental results of (**a**) the PV voltage and output voltage of a boost converter, and (**b**) the duty cycle.

As can be seen in Figures 17 and 18, the converter reacts to the load by decreasing the current from 12.3 A to 5.65 A. The power remains constant at 1030 W for the input and output of the boost converter. This demonstrates that the proposed control system presents a good dynamic response in both grid-connected and islanded modes.

At time $t = 1$ s, the grid is disconnected, and the PV system is subjected to a fault that occurs at the PCC point. At this stage, it is necessary to satisfy the load demand by using the PV system in the islanded mode. The prime concern would be to disconnect the grid and use the available PV system power. Figures 19 and 20 are included to show, respectively, the voltage waveform during the fault and the active and reactive power supplied to both the linear and nonlinear loads.

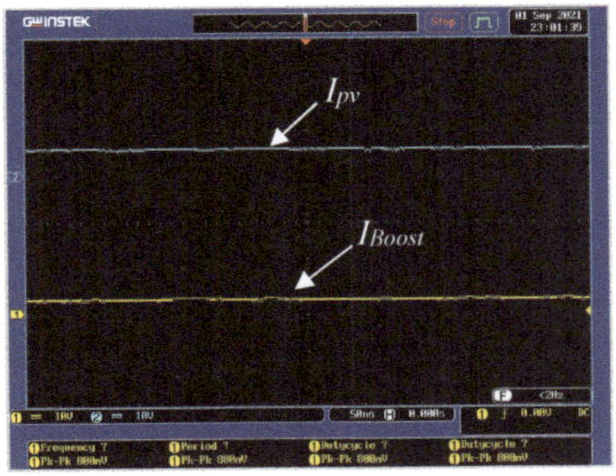


**Figure 17.** The PV and output currents of the boost converter.

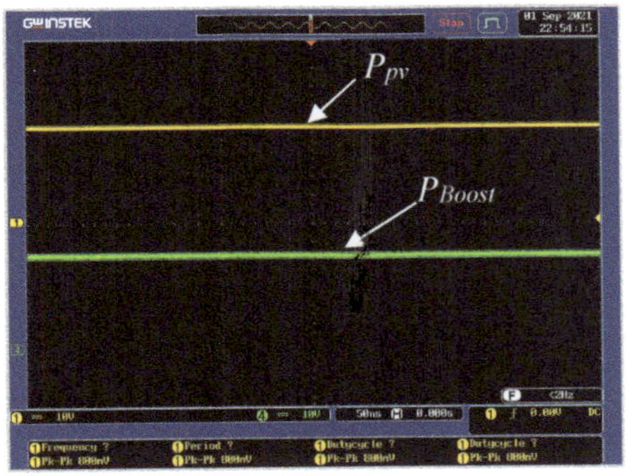


**Figure 18.** The input and output power of the boost converter.

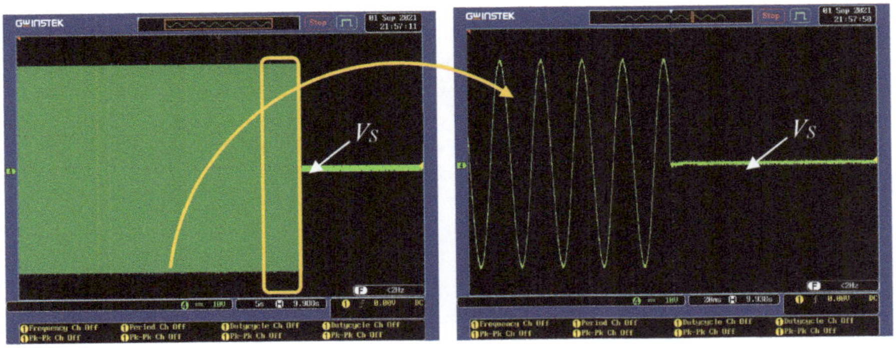


**Figure 19.** The source voltage waveform at the onset of the fault.

Moreover, and as can be seen in Figure 21a, the signals given to the gates of IGBTs 1 and 3 are in opposition to the signals delivered to the gates of IGBTs 2 and 4. Between the pulses delivered to the same branch of the inverter bridge, a dead time of 4 μs between the pulses has been maintained to prevent short circuits, as clearly seen in Figure 21b.

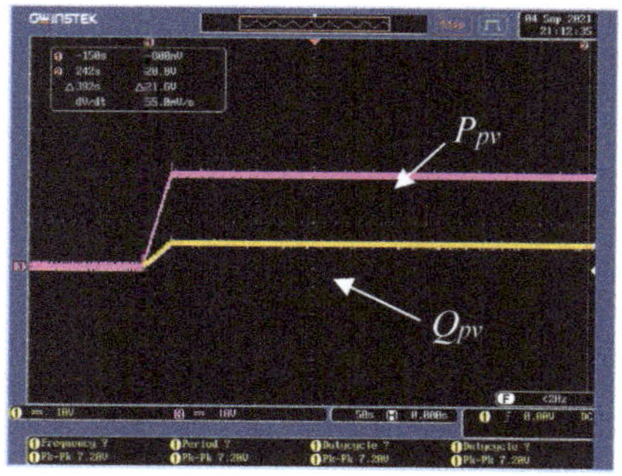


**Figure 20.** Active and reactive power supplied during islanding.

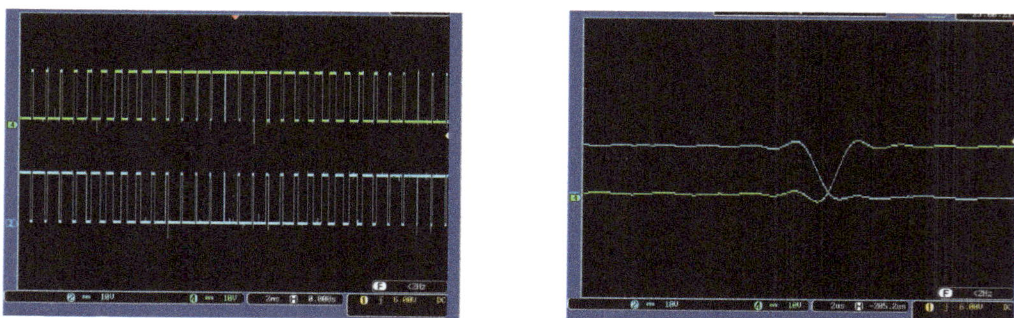

(a) (b)

**Figure 21.** (**a**) The control signal of the switches; (**b**) dead time for two control signals from two switches belonging to the same inverter arm.

*8.2. Load Mode as Supplied Solely by the Grid*

Figures 22 and 23 show the responses of the active and reactive powers supplied to the linear and non-linear loads, respectively. It can be seen that the value of the active power supplied to the linear load is 450 W, and that of the reactive power is equal to 125 Var. For the nonlinear load, the values of (P) and (Q) are 850 W and 210 Var, respectively. This clearly shows that the control system presents good dynamics.

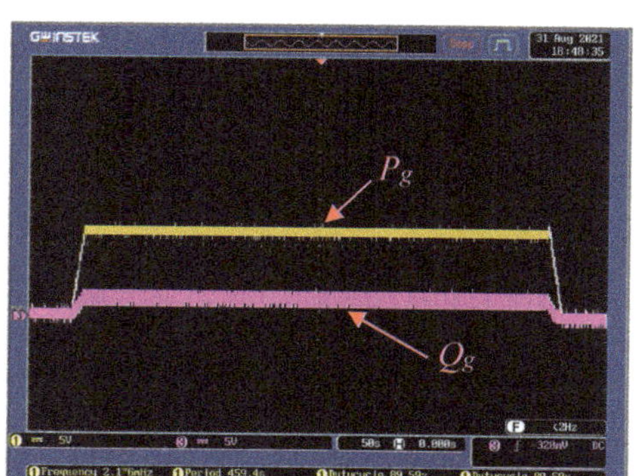


**Figure 22.** Experimental waveforms of the active and reactive power of a linear load.

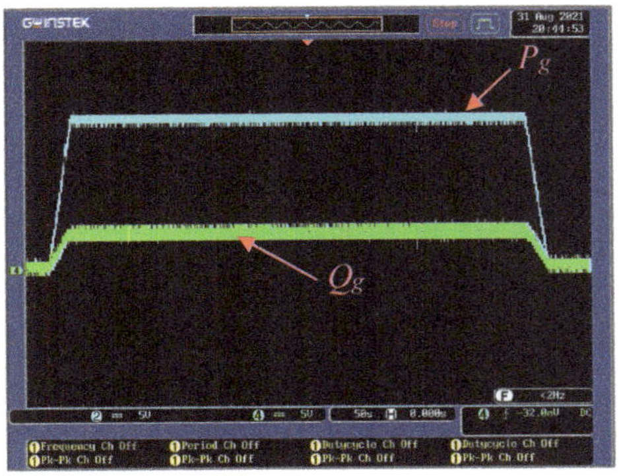


**Figure 23.** Experimental waveforms of the active and reactive power of a non-linear load.

*8.3. Load Mode as Supplied by the Grid and PV System*

When the experimental system is connected to the grid, the source begins to supply the load with sufficient power, as can be seen at the start of Figure 24, where the values of the active and reactive powers are equal to 850 W and 125 Var, respectively. When a fault occurs at the PCC point, the source fails to meet the load demand; hence, the PV system starts to supply power to the load. The corresponding results are clearly shown in Figure 24, where the values of P and Q become 1030 W and 210 Var, respectively.

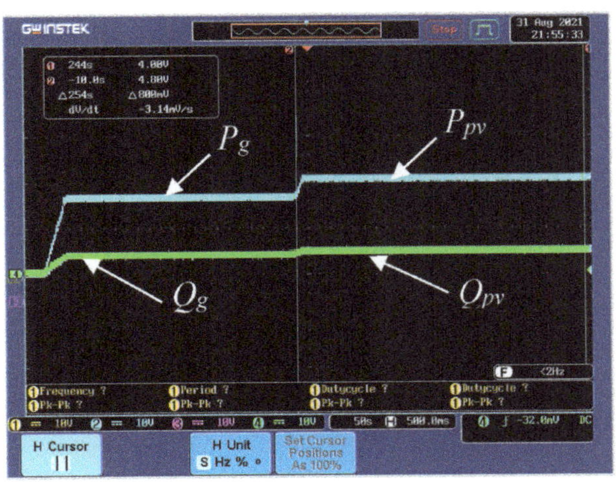


**Figure 24.** Experimental results of the load power from both the source and the PV system.

Figures 25 and 26 show the load current before and after harmonics correction.

The waveform of the load current before correcting the harmonics is presented in Figure 26. The spectral analysis indicates that the THD is 40.8%, as illustrated in Figure 27, where the load is supplied solely by the source. However, after the fault, the THD has reduced drastically to 2.8%, as can clearly be seen in Figure 28.

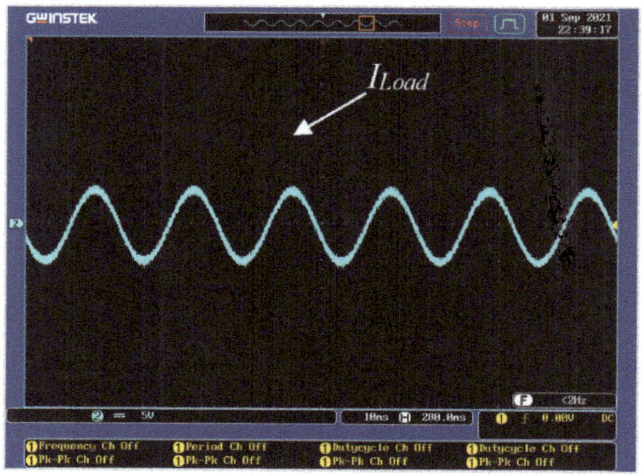


**Figure 25.** Experimental waveforms of the load current after correcting the harmonics.

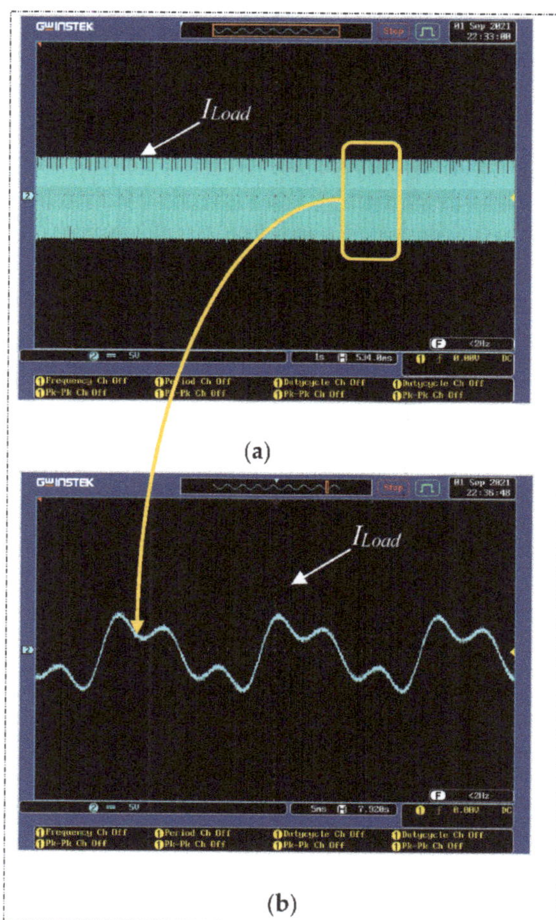


**Figure 26.** Experimental waveforms of the load current before correcting the harmonics. (**a**) Actual load current waveform; (**b**) Zooming of the load current.

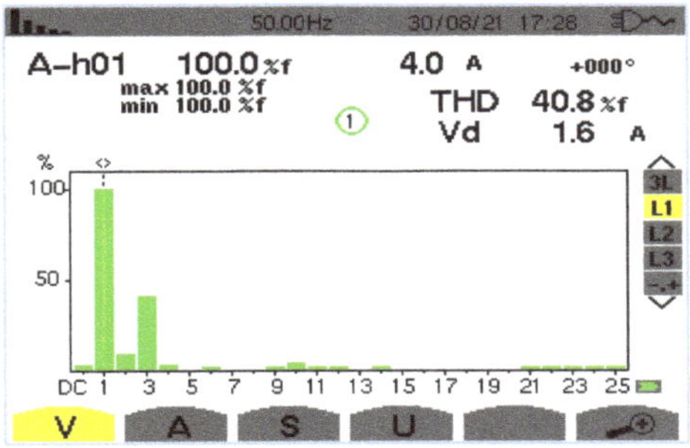


Figure 27. THD contents at the source supply only.

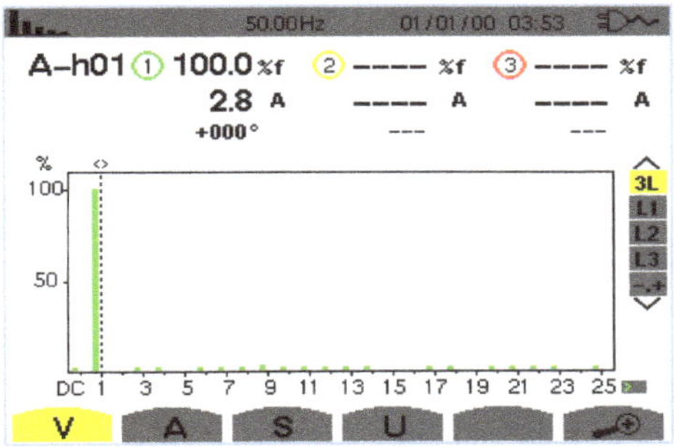


Figure 28. THD contents when the PV system is switched on.

## 8.4. Voltage Sag Mode

During the appearance of the voltage sag after the occurrence of a fault, the system is shifted to sag fault mode, and the current controller still supplies reactive current to restore the voltage according to the grid requirements, as shown in Figure 29.

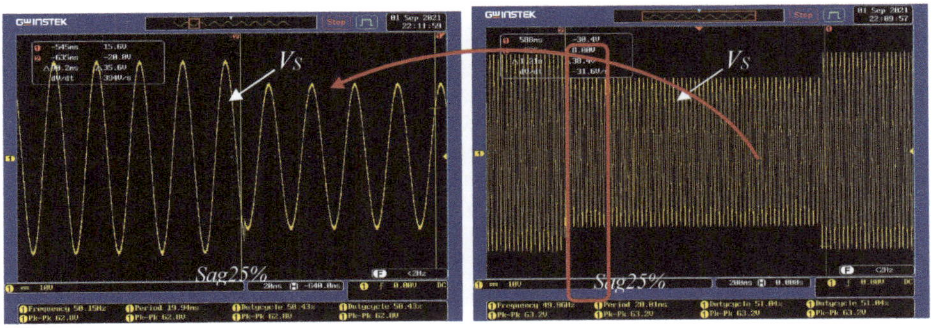


Figure 29. Source voltage during sag.

It is clear from Figure 30 that upon the occurrence of the sag, the load voltage remains unchanged, and the supply to the load is maintained by the PV system, due to the compensating current.

Both Figures 31 and 32 are included to show, respectively, the responses of the active and reactive powers during this sag disturbance and the DC link voltage across the capacitor, which is maintained to be constant at 330 V.

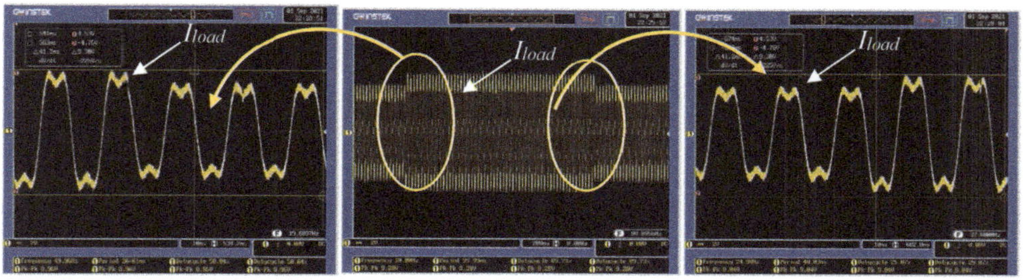


**Figure 30.** The current supplied by the current controller.

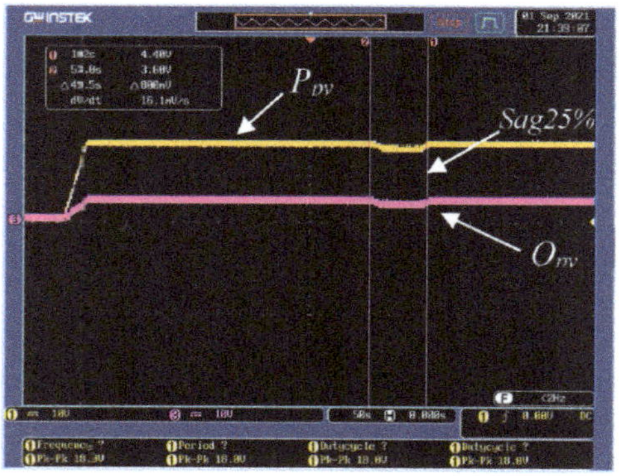


**Figure 31.** The response of the power waveforms during the voltage sag.

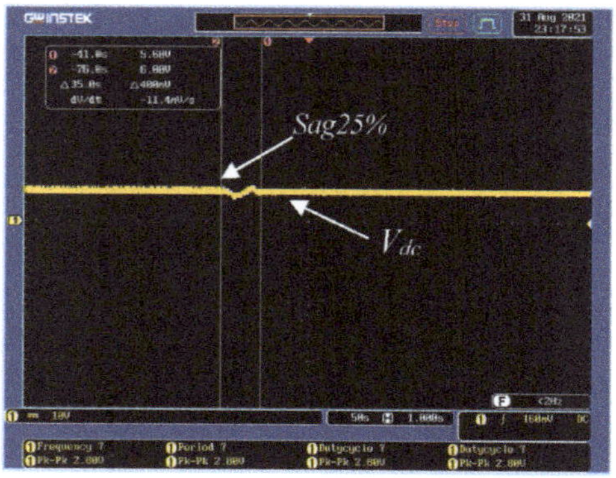


**Figure 32.** The response of the DC voltage waveform during the voltage sag.

We can see the close similarity of the simulated results with those of the experimental tests, as summarized below:

1. Waveforms of real and reactive power are depicted during islanding and during sag voltage, as shown by the simulated results in Figure 9 and those of the experimental test in Figure 20.
2. Waveforms of the DC link voltage and real and reactive power for a linear load are shown in the simulated results in Figure 10 and those of the experimental test in Figure 22.

3. Waveforms of the load supplied by the source and PV as well as the powers for linear and nonlinear loads are shown in the simulated results in Figure 11 and those of the experimental test in Figure 24.
4. Waveforms of the source voltage in fault conditions and during sag voltage are shown in the simulated results in Figure 12 and those of the experimental test in Figures 19 and 29.
5. Finally, the waveforms of the grid current THD analysis are shown in the simulated results in Figure 13 and those of the experimental test in Figures 27 and 28.

## 9. Conclusions

In this paper, a robust and effective control technique for a grid-connected single-phase photovoltaic (PV) system to improve the power quality at the PCC is proposed. The control strategy ensures that the PV inverter manages and performs its functions simultaneously (active power injection, reactive power compensation, and current harmonics filtering) without overrating by limiting its output current. Various scenarios, variable reactive power baselines, and nonlinear load power demands below and above PV power generation have been studied to assess the efficiency of the proposed strategy. Moreover, a performance analysis of the proposed control strategy when applied to a grid-connected inverter powered by PV has been carried out through simulations using MATLAB; some of the simulation results are included in this paper. The control scheme has been validated experimentally on a laboratory-scale PV system and tested under various grid fault and variable load conditions; the main results are included in this paper. The experimental results demonstrate that the proposed control strategy is robust, stable, and efficient. Moreover, in this research, the proposed controller was able to successfully deal with both linear and nonlinear load conditions and demonstrates that the use of the PV system significantly enhances the power quality.

Referencing several studies that have been reviewed, Table 2 summarizes the previous research work investigating different control techniques for grid-connected PV systems, with a brief outline of their merits and disadvantages.

**Table 2.** Summary of the research work related to grid-connected PV system control strategies.

| Ref. | Year | Method | Merits/Disadvantages |
|---|---|---|---|
| [41] | 2017 | Conventional PQ control method | - Ease of implementing the PI controller<br>- Response is within an acceptable range |
| [42] | 2015 | CMPN-based PI controller | - Quick response<br>- Reduced oscillations in power waveforms<br>- Optimum active current is injected during fault time |
| [43] | 2016 | Resonant controller | - Reduced oscillations in DC link voltage<br>- High over current conditions |
| [34] | 2014 | Dynamic resistor braking | - Ease of implementation<br>- No reactive current injection |
| [35] | 2016 | Quasi Z-source inverter | - Regulation of DC link voltage<br>- Reactive power is not considered |
| [44] | 2015 | Energy storage system in supercapacitor | - Ease of design and implementation<br>- Oscillations in power observed |
| [36] | 2019 | Conventional PI-based PQ control | - Easy to design and implement<br>- Transfer of maximum active power into the grid during faults |
| [40] | 2018 | Model predictive control approach | - Does not support over current reduction in the grid<br>- Quick response time |
| [45] | 2017 | Fault ride in control based on fuzzy logic | - Response is rapid and hardware implementation is easy<br>- High over current conditions |
| [46] | 2015 | Probabilistic wavelet fuzzy-based neural network controller | - Requires complex calculation<br>- High over current conditions |

**Author Contributions:** Methodology, A.B. and M.B.S.; Software, A.B.; Validation, A.D.; Formal analysis, A.D. and A.A.; Investigation, A.A. and M.B.S.; Resources, A.D., T.T. and O.A.; Data curation, A.D. and O.A.; Writing—original draft, A.D., T.T. and O.A.; Writing—review & editing, M.D. and M.B.S.; Supervision, M.D. and M.B.S.; Project administration, M.B.S.; Funding acquisition, A.D. All authors have read and agreed to the published version of the manuscript.

**Funding:** This research received no external funding.

**Data Availability Statement:** Not applicable.

**Acknowledgments:** This project was funded by the Deanship of Scientific Research (DSR) at the Islamic University of Madinah (IUM) in Saudi Arabia, under the Institutional Finance Program "Tamayyuz 2" via Grant No. 656 in November 2020. The authors, therefore, acknowledge with gratitude the support of DSR in IUM and wish to extend their appreciation for the full financial and moral support received.

**Conflicts of Interest:** The authors declare that there is no conflict of interest.

## Abbreviations

| | |
|---|---|
| PV | Photovoltaic |
| PWM | Pulse width modulation |
| MPPT | Maximum power point tracking |
| ANFIS | Adaptive neuro-fuzzy inference |
| OSG | Orthogonal signal generator |
| THD | Total harmonic distortion |
| PCC | Point of common coupling |
| PQ | Power quality |
| PLL | phase-locked loops |
| LP | Low pass filter |
| FRT | Fault ride-through |
| P&O | Perturb and observe |
| $I_{PV}$, $V_{PV}$ | PV array current and voltage |
| $I_{gE}$, $I_{gF}$ and $V_{gE}$, $V_{gF}$ | Grid current and voltage |
| $P$ $Q$ | The real and reactive powers |
| $h$ | Harmonic order |
| $\omega_0$ | Fundamental frequency |
| $T_s$ | Sampling period |
| $P_{MPP}$ | PV array MPP power |
| N | strings of PV |

## Appendix A

The proposed system uses the ANFIS system, with E and CE as the inputs and y as the output. The fuzzy IF-THEN rules are implemented as follows in ANFIS:

Rule1:
IF 'E' is $E_1$ and 'CE' is $EC_1$ Then $y = a_1E + b_1CE + c_1$
Rule1:
IF 'E' is $E_2$ and 'CE' is $EC_2$ Then $y = a_2E + b_2CE + c_2$

In Figure 6, the ANFIS controller's Takagi–Sugeno model-based design on five layers is shown. The function of each layer is given below:

L1: This layer receives the inputs and converts them to fuzzy sets. Each node '$i$' has a function that allows it to be flexible.

$$O_{1,I} = \mu_{Mi}(\text{E}) \text{ for } i = 1,2 \quad \text{(A1)}$$

$$O_{1,I} = \mu_{Ni-2}(\text{E}) \text{ for } i = 3,4 \quad \text{(A2)}$$

The inputs are E and CE, and the linguistic labels are Mi and Ni-2. The triangle MFs are utilized and the resulting MFs once training is finished are displayed in Figure 10.

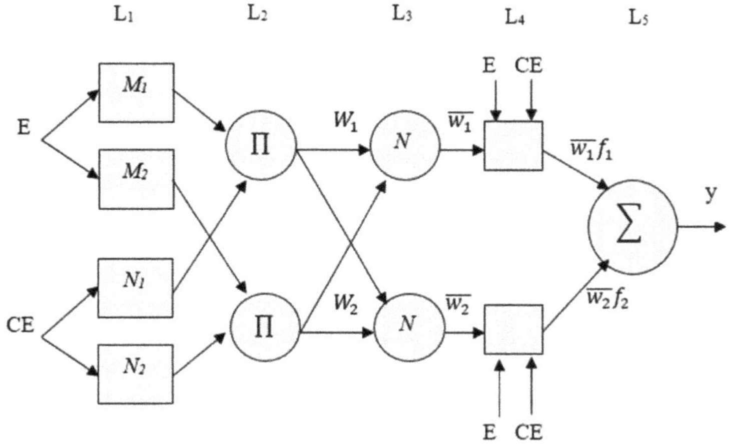


**Figure A1.** The architecture of ANFIS.

L2: The multiplication of all incoming signals is one of this layer's outputs. This layer's nodes are all static $\pi$ nodes.

$$O_{2,I} = w_i = \mu_{Mi}(\text{E})\mu_{Ni}(\text{CE}) \quad \text{(A3)}$$

The rule firing strength is represented by a node output. The fuzzy AND is used as a function.

L3: Normalization of the rules is calculated in this layer. It is the ratio of the firing strength of each rule to the total value of the firing strengths. The nodes are labeled N and are static.

$$O_{3,\,i} = \overline{w_i} = \frac{w_i}{w_1 + w_2}\; i = 1, 2, \ldots \quad \text{(A4)}$$

L4: This is the defuzzification layer, where the nodes are versatile by a function.

Here, $\overline{w_i}$ is the normalized firing strength and the values of $a_1$, $b_1$, and $c_1$ are the subsequent parameters.

L5: This contains a static node, labeled Σ, which will add together all incoming signals to calculate the output.

$$O_{5,i} = y = \sum \overline{w_i} f_i = \frac{\sum w_i f_i}{\sum w_i} \quad \text{(A5)}$$

All outputs were set to zero before the ANFIS system began training. Following the completion of training, the ANFIS system produced output values, which are listed in Table A1. The ANFIS system was educated using the MATLAB platform's hybrid learning method. The various datasets seen in the experimental results using the traditional controller were used to train the ANFIS system to enhance its performance. Figure 11 depicts the ANFIS controller's surface view. In a three-dimensional perspective, it displays the connection between two inputs and outputs. It aids in comprehending the controller output's performance as a function of the inputs.

**Table A1.** Training parameter values for the ANFIS system.

| Nomenclature | Value |
|---|---|
| TDPs | 350,000 |
| CDP | 60,000 |
| MF | 7 |
| Nodal points | 131 |
| NP | 42 |
| LP | 49 |
| TP | 91 |
| FR | 49 |
| NE | 150 |

## Appendix B

**Table A2.** Output values of the ANFIS system.

| CE<br>E | $EC_1$ | $EC_2$ | $EC_3$ | $EC_4$ | $EC_5$ | $EC_6$ | $EC_7$ |
|---|---|---|---|---|---|---|---|
| $E_1$ | −0.5 | −0.3 | −0.0030 | 0 | 0 | 0 | 0 |
| $E_2$ | −0.35 | −0.14 | 0.07 | 0 | 0 | 0 | 0 |
| $E_3$ | 0 | 0.07 | 0.28 | 0.42 | −0.02 | 0 | 0 |
| $E_4$ | 0 | 0 | 0.5 | 0.7 | 0.5 | −0.06 | 0 |
| $E_5$ | 0 | 0 | 0 | 1.2 | 0.9 | 1.06 | 0.01 |
| $E_6$ | 0 | 0 | 0 | 0 | 0.9 | 0.9 | 1.3 |
| $E_7$ | 0 | 1.1 | 0.6 | 0 | 0 | 0.6 | 0.9 |

## Appendix C

Details of the Simulink implementation of a single-phase grid-connected PV system with an LCL filter, as depicted in Figure A2.

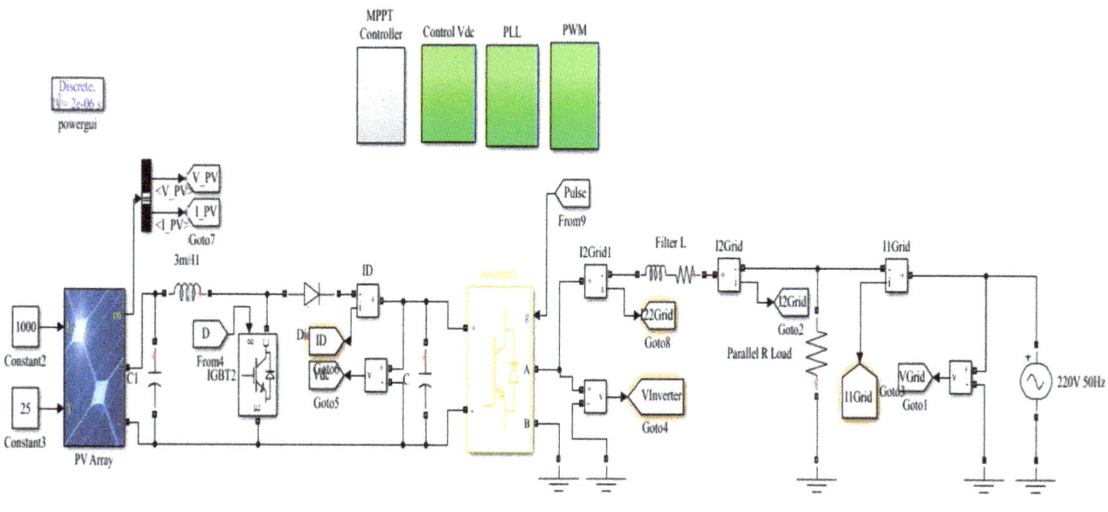


**Figure A2.** Simulink model of the single-phase grid-connected PV system.

## References


1. Dugan, R.C.; Mc Granaghan, M.F.; Beaty, H.W. *Electrical Power Systems Quality*; McGraw-Hill: New York, NY, USA, 1996.
2. Farhadi-Kangarlu, M.; Babaei, E.; Blaabjerg, F. A comprehensive review of dynamic voltage restorers. *Int. J. Electr. Power Energy Syst.* **2017**, *92*, 136–155. [CrossRef]
3. Sankaren, C. *Power Quality*; CRC Press: Boca Raton, FL, USA, 2017.
4. *IEEE Std. 1346-1998*; IEEE Recommended Practice for Evaluating Electric Power System Compatibility with Electronic Process Equipment. IEEE: Piscataway, NJ, USA, 1998.
5. *IEC Std. 61000-2-1*; Electromagnetic Compatibility (EMC), Part 2: Environment, Section 1: Description of the Environment, Electromagnetic Environment for Low-Frequency Conducted Disturbances and Signaling in Public Power Supply Systems. IEC: Geneva, Switzerland, 1990.
6. Rauf, A.M.; Khadkikar, V. Integrated photovoltaic and dynamic voltage restorer system configuration. *IEEE Trans. Sustain. Energy* **2015**, *6*, 400–410. [CrossRef]
7. Kusko, A. *Power Quality in Electrical Systems*; McGraw-Hill Education: New York, NY, USA, 2007.
8. Gotekar, P.S.; Muley, S.P.; Kothari, D.P. A Single Phase Grid Connected PV System working in Different Modes. *Eng. Technol. Appl. Sci. Res.* **2020**, *10*, 6374–6379. [CrossRef]
9. Proceedings of the 27th Conference of the Parties to the United Nations Framework Convention on Climate Change, COP27, Sharm el-Sheikh, Egypt, 6–20 November 2022; Available online: https://www.un.org/en/climatechange/cop27 (accessed on 8 May 2023).
10. Shukhobodskiy, A.A.; Colantuono, G. RED WoLF: Combining a battery and thermal energy reservoirs as a hybrid storage system. *Appl. Energy* **2020**, *274*, 115209. [CrossRef]
11. Toosi, H.A.; Del Pero, C.; Leonforte, F.; Lavagna, M.; Aste, N. Machine learning for performance prediction in smart buildings: Photovoltaic self-consumption and life cycle cost optimization. *Appl. Energy* **2023**, *334*, 120648. [CrossRef]
12. Teodorescu, R.; Liserre, M.; Rodríguez, P. *Grid Converters for Photovoltaic and Wind Power Systems*; John Wiley & Sons: Chichester, UK, 2011.
13. Mahamudul, H.; Saad, M.; Henk, M.I. Photovoltaic system modeling with fuzzy logic based maximum power point tracking algorithm. *Int. J. Photoenergy* **2013**, *2013*, 762946. [CrossRef]
14. Sangwongwanich, A.; Blaabjerg, F. Mitigation of interharmonics in PV systems with maximum power point tracking modification. *IEEE Trans. Power Electron.* **2019**, *34*, 8279–8282. [CrossRef]
15. Gotekar, P.S.; Muley, S.P.; Kothari, D.P. Fuzzy Inference based control of single phase grid connected photovoltaic system. In Proceedings of the 2019 Innovations in Power and Advanced Computing Technologies (i-PACT), Vellore, India, 22–23 March 2019; pp. 1–3.
16. Vosoughi, N.; Hosseini, S.H.; Sabahi, M. Single-phase common-grounded transformer-less grid-tied inverter for PV application. *IET Power Electron.* **2020**, *13*, 157–167. [CrossRef]
17. Ghosh, S.S.; Nathan, K.S.; Siwakoti, Y.P.; Long, T. Dual polarity DC–DC converter integrated grid-tied single-phase transformer less inverter for solar application. *J. Eng.* **2019**, *2019*, 3962–3966. [CrossRef]
18. Datta, A.; Sarker, R.; Hazarika, I. An efficient technique using modified p–q theory for controlling power flow in a single-stage single-phase grid-connected PV system. *IEEE Trans. Ind. Inform.* **2018**, *15*, 4635–4645. [CrossRef]
19. Anurag, A.; Yang, Y.; Blaabjerg, F. Thermal performance and reliability analysis of single-phase PV inverters with reactive power injection outside feed-in operating hours. *IEEE J. Emerg. Sel. Top. Power Electron.* **2015**, *3*, 870–880. [CrossRef]
20. Yang, Y.; Wang, H.; Blaabjerg, F. Reactive power injection strategies for single-phase photovoltaic systems considering grid requirements. *IEEE Trans. Ind. Appl.* **2014**, *50*, 4065–4076. [CrossRef]
21. Yang, Y.; Blaabjerg, F.; Zou, Z. Benchmarking of grid fault modes in single-phase grid-connected photovoltaic systems. *IEEE Trans. Ind. Appl.* **2013**, *49*, 2167–2176. [CrossRef]
22. Yang, Y.; Blaabjerg, F. Low-voltage ride-through capability of a single-stage single-phase photovoltaic system connected to the low-voltage grid. *Int. J. Photoenergy* **2013**, *2013*, 257487. [CrossRef]
23. Ikken, N.; Bouknadel, A.; Haddou, A.; Tariba, N.-E.; El Omari, H.; El Omari, H. PLL synchronization method based on second-order generalized integrator for single phase grid connected inverters systems during grid abnormalities. In Proceedings of the 2019 International Conference on Wireless Technologies, Embedded and Intelligent Systems (WITS), Fez, Morocco, 3–4 April 2019; pp. 1–5.
24. Butt, H.Z.; Awon, M.; Khalid, H.A. Performance Analysis of a Continuous and Discretized Second Order Generalized Integrator based Phase Lock Loop for Single Phase Grid Connected PV Systems. In Proceedings of the 2018 International Conference on Power Generation Systems and Renewable Energy Technologies (PGSRET), Islamabad, Pakistan, 10–12 September 2018; pp. 1–6.
25. Panda, A.K.; Patnaik, N. Combined operation of a new power angle control unit vector template based unified power quality conditioner and fuel cell stack supply with effective utilization of shunt and series inverter. *Electr. Power Compon. Syst.* **2016**, *44*, 2048–2058. [CrossRef]
26. Nouaiti, A.; Mesbahi, A.; Saad, A.; Khafallah, M.; Reddak, M. Realization of a Single-Phase Multilevel Inverter for Grid-Connected Photovoltaic System. *Eng. Technol. Appl. Sci. Res.* **2018**, *8*, 3344–3349. [CrossRef]
27. Krithiga, S.; Gounden, N.G.A. Power electronic configuration for the operation of PV system in combined grid-connected and stand-alone modes. *IET Power Electron.* **2014**, *7*, 640–647. [CrossRef]

28. Subudhi, P.S.; Krithiga, S. PV and grid interfaced plug-in EV battery charger operating in P-VG PV and VG modes. *Int. J. Recent Technol. Eng.* **2019**, *8*, 3431–3443. [CrossRef]
29. Reddy, V.R.; Sreeraj, E.S. A feedback-based passive islanding detection technique for one-cycle-controlled single-phase inverter used in photovoltaic systems. *IEEE Trans. Ind. Electron.* **2019**, *67*, 6541–6549. [CrossRef]
30. Raza, S.A.; Jiang, J. A benchmark distribution system for investigation of residential microgrids with multiple local generation and storage devices. *IEEE Open Access J. Power Energy* **2019**, *7*, 41–50. [CrossRef]
31. Lam, L.H.; Phuc, T.D.H.; Hieu, N.H. Simulation models for three-phase grid connected PV inverters enabling current limitation under unbalanced faults. *Eng. Technol. Appl. Sci. Res.* **2020**, *10*, 5396–5401. [CrossRef]
32. Jamil, M.; Rizwan, M.; Kothari, D.P. *Grid Integration of Solar Photovoltaic Systems*; CRC Press: Boca Raton, FL, USA, 2017.
33. Kazmierkowski, M.P. Renewable Energy Devices and Systems with Simulations in MATLAB and ANSYS [Book News]. *IEEE Ind. Electron. Mag.* **2018**, *12*, 80–83. [CrossRef]
34. Hasanien, H.M. An adaptive control strategy for low voltage ride through capability enhancement of grid-connected photovoltaic power plants. *IEEE Trans. Power Syst.* **2015**, *31*, 3230–3237. [CrossRef]
35. Cárdenas, R.; Díaz, M.; Rojas, F.; Clare, J.; Wheeler, P. Resonant control system for low-voltage ride-through in wind energy conversion systems. *IET Power Electron.* **2016**, *9*, 1297–1305. [CrossRef]
36. Ben Abdelkader, A.; Toumi, T.; Abdelkhalek, O. Experimental verification of dynamic voltage restorer fed by solar PV: Lithium-ion battery storage for lasting power quality improvement. *Electr. Eng.* **2022**, *104*, 4581–4593. [CrossRef]
37. Hossain, M.K.; Ali, M.H. Fuzzy logic controlled power balancing for low voltage ride-through capability enhancement of large-scale grid-connected PV plants. In Proceedings of the 2017 IEEE Texas Power and Energy Conference (TPEC), College Station, TX, USA, 9–10 February 2017; pp. 1–6.
38. Lin, F.-J.; Lu, K.-C.; Ke, T.-H.; Yang, B.-H.; Chang, Y.-R. Reactive power control of three-phase grid-connected PV system during grid faults using Takagi–Sugeno–Kang probabilistic fuzzy neural network control. *IEEE Trans. Ind. Electron.* **2015**, *62*, 5516–5528. [CrossRef]
39. Toumi, T.; Allali, A.; Meftouhi, A.; Abdelkhalek, O.; Benabdelkader, A.; Denai, M. Robust control of series active power filters for power quality enhancement in distribution grids: Simulation and experimental validation. *ISA Trans.* **2020**, *107*, 350–359. [CrossRef]
40. Al-Durra, A.; Fayyad, Y.; Muyeen, S.; Blaabjerg, F. Fault ride-through of a grid-connected photovoltaic system with quasi Z source inverter. *Electr. Power Compon. Syst.* **2016**, *44*, 1786–1800. [CrossRef]
41. Bae, Y.; Vu, T.-K.; Kim, R.-Y. Implemental control strategy for grid stabilization of grid-connected PV system based on German grid code in symmetrical low-to-medium voltage network. *IEEE Trans. Energy Convers.* **2013**, *28*, 619–631. [CrossRef]
42. Blaabjerg, F.; Teodorescu, R.; Liserre, M.; Timbus, A.V. Overview of control and grid synchronization for distributed power generation systems. *IEEE Trans. Ind. Electron.* **2006**, *53*, 1398–1409. [CrossRef]
43. Hasaneen, M.M.; Badr, M.A.; Atallah, A.M. Control of active/reactive power and low-voltage ride through for 40 kW three-phase grid-connected single-stage PV system. *CIRED-Open Access Proc. J.* **2017**, *2017*, 1655–1659. [CrossRef]
44. Hossain, M.K.; Ali, M.H. Low voltage ride through capability enhancement of grid connected PV system by SDBR. In Proceedings of the 2014 IEEE PES T&D Conference and Exposition, Medellin, Colombia, 10–13 September 2014; pp. 1–5.
45. Worku, M.Y.; Abido, M.A. Grid-connected PV array with supercapacitor energy storage system for fault ride through. In Proceedings of the 2015 IEEE International Conference on Industrial Technology (ICIT), Seville, Spain, 17–19 March 2015; pp. 2901–2906.
46. Bighash, E.Z.; Sadeghzadeh, S.M.; Ebrahimzadeh, E.; Blaabjerg, F. Improving performance of LVRT capability in single-phase grid-tied PV inverters by a model-predictive controller. *Int. J. Electr. Power Energy Syst.* **2018**, *98*, 176–188. [CrossRef]

energies MDPI

Article

# Governance and Performance in Romanian Energy Companies

Adrian Doru Bigioi [1,*] and Cristina Elena Bigioi [2]

1 Department of Accounting and Audit, The Faculty of Accounting and Management Information Systems, The Bucharest University of Economic Studies, 010374 Bucharest, Romania
2 Department of Economics, Faculty of Entrepreneurship, Business Engineering and Management, University Politehnica of Bucharest, 060042 Bucharest, Romania; cristina.dumitru@upb.ro
* Correspondence: adrian.bigioi@cig.ase.ro

**Abstract:** The energy sector is essential to a country's economy and social life. In recent years, the energy market in Romania has experienced a certain dynamism, influenced by multiple factors such as regional conflicts, national and EU energy legislation, liberalization of the energy market, and increasing demand for energy products. These factors determined a sharp increase in electricity, gas, and oil prices in Romania in recent years. Based on the above, the present study aims to analyze Romanian energy companies' financial performances and determine whether there is any correlation between their governance and their performance. In order to fulfill the research objectives, financial performance data were extracted and analyzed for a number of 43 energy companies in Romania, considering the financial statements prepared up to 31 December 2021. Using the scoring method, governance indices were determined for each company. Finally, using a linear regression model, we analyzed the correlation between governance and performance of the 43 companies. The results of the study indicate that there is a direct correlation of moderate intensity between the two variables, with the performance of energy companies being higher if they have implemented a governance system.

**Keywords:** energy companies; financial performance; corporate governance

## 1. Introduction

Corporate governance is defined as the system by which companies are governed and controlled [1]. The professional body IFAC gives several definitions of corporate governance concepts. Thus, corporate governance involves a set of relationships between a company's executive management, board of directors, shareholders, and other associated parties. Corporate governance provides the structure through which the objectives of the firm are set, as well as the means of achieving those objectives and monitoring performance. Good corporate governance should provide the motivation for achieving objectives that are in the best interests of the firm and its shareholders and should facilitate effective monitoring, thereby encouraging firms to use resources efficiently [2].

According to the IFAC recommendations, corporate governance is a conceptual framework for organizations, with two dimensions: compliance and performance. Compliance covers the role and structure of management as well as its remuneration. The performance focuses on strategy and value creation. This dimension helps management to make strategic decisions, identify risks, find the key to performance, and identify key decision points. Therefore, there is a need to develop best practices and techniques to be used effectively in different economic entities [3].

Internationally, the concept was pioneered by the UK private sector, i.e., by shareholders who wanted better control over the management of companies while maximizing financial results. Subsequently, the concept was taken up and implemented in several countries, thanks to the role played by the Organization for Economic Co-operation and Development (OECD) in publishing generally applicable principles of a recommendatory nature. The first set of principles was developed in 1999, and an improved version was published in 2004 [1].

The O.E.C.D. has also published a series of guidelines on the implementation of these principles by country states at a national level. These principles cover: shareholders' rights (guarantee of ownership, right to dividends, right to be informed of important decisions affecting the company's business, right to participate and vote in general meetings), fair treatment of all shareholders (minority shareholders enjoying the same rights as majority shareholders), the role of the different parties involved in the management and control of the company (employees, creditors, suppliers, customers, financial institutions), transparency and disclosure of information (the corporate governance system ensures timely and accurate information to all stakeholders on the financial statements, results, shareholding and governance of the company), the responsibilities of the board of directors and executive management for their activities.

These principles have been incorporated in several scientific papers, with most studies showing that implementing corporate governance principles at company levels has beneficial effects for shareholders, leading to the maximization of their profits. In recent years, however, the world economy has faced and is still facing financial problems caused by several factors, and, for this reason, it has been questioned whether better corporate governance can lead to greater global economic and financial stability.

The corporate governance system is primarily designed to protect the interests of investors. The losses caused to shareholders by the management of multinational companies because of financial scandals have led some of them to develop a mechanism to protect their interests.

Adrian Cadbury (1992) is one of the promoters of the corporate governance concept. The concept has also been adopted to a certain extent at the international level, with the management of multinational companies being responsible for implementing the principles of corporate governance at the level of the entities' activities [4].

Internationally, there are several corporate governance models, including the shareholder model and the stakeholder model. The shareholder model is the model found in Anglo-Saxon countries. It is characterized by the fact that its implementation in companies is aimed at maximizing the economic benefits of shareholders, profits and share prices being among them. The stakeholder model is present in most European countries except the Anglo-Saxon and Nordic countries. It is characterized by the fact that its implementation in companies is not necessarily aimed at maximizing the economic advantages of shareholders but at protecting the interests of all parties involved in the corporate governance mechanism (customers, suppliers, employees, shareholders, managers, and state institutions) [5].

Inadequate governance and control of public interest enterprises can affect global economic and financial stability, in some cases leading to global financial crises. Based on these considerations, the most developed countries issued the "OECD Principles of Corporate Governance" in 1999.

These principles are one of the basic pillars of global economic and financial stability. Developed countries have implemented these corporate governance principles in their legislation. The US, for example, in 2002, hit by financial scandals, passed the Sarbanes-Oxley Act, which introduced certain governance obligations for US companies. European Union countries have also introduced corporate governance principles into their legislation. The United Kingdom has put shareholders at the forefront, and countries such as France, Germany, and Italy have paid more attention to the role of employees and financial institutions as participants in corporate governance.

In view of the above-mentioned role of governance in maximizing the financial performance of companies and the fact that electricity, gas, and oil are indispensable in economic and social life, this study aims to analyze the degree of implementation of governance principles in Romanian energy companies.

Additionally, given that energy entities can be considered of public interest, contributing to ensuring macroeconomic balance, we consider it necessary that they present financial stability. Starting from the premise that the implementation of corporate governance or its improvement can influence the efficiency of the activity of energy companies and increase their financial performance, this study aims to analyze the correlation between governance and performance in Romanian energy companies. The present study is important because it answers the question of whether better corporate governance can prevent the decline in the financial results of energy companies, their default, and possible national and even international energy and economic crises. The results of the study can then be used by companies to improve their corporate governance and risk management strategies. Good corporate governance is particularly important in financial reporting, as it is a key factor in ensuring confidence in the capital markets by providing quality information. Good corporate governance emphasizes the importance of non-executive directors, the audit committee, and their structures and relationships with management. It also focuses on internal controls, internal audits, external audits, and corporate governance disclosures.

The energy sector is essential to a country's economy and social life. In recent years, the energy market in Romania has experienced a certain dynamism, influenced by factors such as regional conflicts, national and EU energy legislation, liberalization of the energy market, and increasing demand for energy products. These factors determined a sharp increase in electricity, gas, and oil prices in Romania in recent years. Below we have presented the evolution of annual inflation of the prices of energy goods in Romania according to the forecasts of the National Bank of Romania (Figure 1).

**Figure 1.** Annual inflation of energy goods prices. Source: processing data from the graph mentioned in: https://www.bnr.ro/Proiectii-BNR-22694-mobile.aspx, accessed on 26 April 2023 [6].

Annual inflation of the prices of energy goods of over 35% is observed in the second quarter of 2022, and, according to the National Bank of Romania's forecasts, the prices of these goods should stabilize in the period of 2023–2024, but we believe that some factors are likely to continue to influence the dynamics of these prices. Additionally, considering the dependence of the companies' activities on these energy products, there was a chain increase in prices in Romania, with inflation being 16.4% in December 2022 compared to December 2021.

The increase in inflation overlapped with the increase in the monetary policy interest rate in December 2022, which was 6.75%. These factors influenced the purchasing power of the population, and, from there, it resulted in a series of chain effects on the demand for goods and services. For example, one of the strongest effects was related to the increase in prices for metallurgical products such as concrete steel used in construction works whose price even exceeded 7 lei per kilogram at the beginning of 2022, the metallurgical market being dependent on the price of electricity used in this industry.

Electricity and gas prices have significantly risen, seeing the highest values in Eurostat records in the second half of the year 2022. Policies and interventions in EU countries have been aimed at mitigating the effects of higher energy sales prices to final customers because of the unprecedented rise in payment costs on the local and European markets. The measures adopted in Romania in this respect were price capping, compensation of energy bills, and temporary tax exemption for consumers [7]. In the European Union countries, the highest increase in electricity prices for household consumers in 2022 compared to 2021 was recorded by Romania (+112%), followed by the Czech Republic (+97%) and Denmark (70%). In terms of gas prices, Romania is the second country with the highest increases (+165%), with the Czech Republic (+231%) in the first place.

Given the importance of energy goods in the economic and social life of a country, we proposed study to analyze the trend in terms of the financial performance of energy companies in this case in Romania in the period 2012–2021 and whether there is any correlation between the governance of these companies and their financial performance.

The paper is organized as follows: next, the literature on the role of corporate governance and performance relationship testing is reviewed and described in Section 2. The research methodology is described in Section 3, followed by a presentation of results (Section 4) and a discussion on the results (Section 5). Research findings, implications, and limitations are discussed in Section 6.

## 2. Literature Review

### 2.1. Importance of Corporate Governance

The COVID-19 crisis has had a significant economic impact worldwide and generated different reactions from different business segments. The crisis has introduced exceptional uncertainty in running a business [8]. Can the adoption of corporate governance diminish the impact of such a crisis?

In times of crisis, several governance mechanisms can help, namely, risk management committees, board diversity, independent directors, foreign investors, institutional ownership, ownership concentration, CEO's dual roles, family ownership [9,10]. The presence of independent commissioners (with monitoring role) can protect the decline of ROA in times of crisis [11].

Additionally, appropriate internal corporate governance can prevent stock price crash risk [12].

The COVID-19 crisis reduced performance and increased risks for private businesses in many sectors. The quality of the country's governance system can prevent the effect of an economic shock (such as the COVID-19 crisis) on firms' operations [13].

Corporate governance mechanisms have an impact on SME perceptions of COVID-19's influence, alerting companies to adjust their business strategy and organizational structure to better cope with the effects of the current crisis [14]. Various studies have examined the characteristics of corporate governance: the difference made by diversity on corporate board [15–17], independence and financial knowledge for the board and the audit committee [18–21], role of auditors [22–24], owner structure [25–28].

Other studies have tested the relationship between the existence of corporate governance and firm performance, demonstrating that corporate governance adds value to firm performance [29–31].

Using the principles of corporate governance positively influences the performance of state-owned enterprises [32–34].

## 2.2. The Energy Sector

Several international research studies have been developed on the concepts of performance and governance in the field of energy companies, and most of them have demonstrated that better corporate governance causes an increase in the economic and financial performance of companies.

Corporate governance practices influence sustainability as part of an organization's corporate social responsibility performance. Boards of directors need to focus on innovation and organizational change to increase sustainability in the social and environmental dimensions and adapt to the new turbulent times [35].

Governance mechanisms, such as environmental management system certification, environmental organization, publication of sustainability reports, strategic environmental planning, and quality of governance, can help combat climate change influenced by carbon emissions from the electricity industry [36].

Financial auditing of energy companies should integrate sustainable development and social responsibility, as currently these companies only follow financial performance that is not significantly affected by social responsibility [37].

The existence of a significant number of independent board members could have a positive impact on environmental performance [38].

Companies' adoption of renewable energy is influenced by both internal governance (indicated by the characteristics of the board of directors) and external governance (indicated by the institutional environment). Adoption increases in the case of a strong external governing body and decreases in the case of the duality of the board of directors and the greater weight of the executive. Firms in common law systems tend to use fewer renewable sources [39].

ESG (environmental, social, and governance) sustainability reporting significantly influences operational performance, especially in emerging economies. The influence is insignificant on financial performance and market performance, according to a study on sectoral energy performance in economies in 50 countries in the period 2008–2017 [40].

Different combinations of ESG pillars (environmental, social, and governance aspects) can influence the financial performances of companies differently. The social pillar improves financial performance [41,42].

Total assets, the amount of proactive environmental costs, the proportion of technicians, and the number of patents contribute most to companies' environmental performances [43,44].

In the communication of companies with stakeholders, an important role is played by the disclosure of non-financial information, especially those related to corporate governance. Disclosure should be mandatory, even for information that is not considered to be in the best interest of companies (such as executive remuneration and employment policy) [45]. Earnings per share, return on equity, type of company, size of company, age of company, and auditing firm positively influence the company environmental disclosure [46].

For companies in the energy sector, performance is influenced by various factors:

- number of board members (a positive influence) [47];
- the presence of women on the board (a negative influence) [47];
- the higher number of directorships of directors (high tenure affects the fulfilment of duties and independence and weakens corporate governance) [48];
- gender diversity of executives [49]: innovation is negatively affected by the proportion of female executives. Economic innovation is promoted by women in high tax burden areas. Women's caution in eco-innovation decisions stems from high investment risk. The results of a study performed for listed companies in China's energy industry demonstrate the importance of CEO–CFO pairing indicators on corporate decision-making and crash risk. It was found that the assignment of a CEO (chief executive

officer) and CFO (chief finance officer) of the same gender had a beneficial impact on the long-term share price stability of energy companies [49].

The positive relationship between corporate governance and company valuation has been demonstrated: the higher the governance standards, the higher their ROA, ROE, or Tobin's Q [50,51].

Corporate governance has a significant impact on the capital structure of companies in the energy industry. The level of leverage is negatively affected by board size and state ownership, two indicators of the quality of corporate governance. The financial expertise of larger boards allows energy companies to finance themselves predominantly through their own capital. Debt financing is negatively influenced by the state as the dominant investor [52].

Governance positively affects operational efficiency. ESG compliance is more pronounced in European companies. Asian companies are more disciplined in the energy sector. Accounting performance (ROA and ROE) differs significantly between GRI (Global Reporting Initiatives) and non-GRI companies [53].

A survey of 55 European companies in different categories in the energy industry indicated a positive association between ESG disclosure and firm value measured using market value [54]. The profitability of electricity companies was positively influenced by vertical integration and was lower for state-owned companies [55].

In Romania, the sustainability of renewable energy companies is questionable. Their investments have been based on state subsidies. Innovation and improvement of business processes allows for improved financial performance [56].

The financial performance of Romanian companies is positively affected by corporate social responsibility actions [57].

The relation between governance and performance was also studied for entities in the banking and insurance sectors in Romania [58,59].

Literature in the field shows that the purpose of corporate governance is to increase the value of a company, and this value is itself influenced by the increase in economic and financial performance (such as turnover). Starting from this, the present study aims to investigate whether the implementation of a governance system in Romanian energy companies has shown to improve their financial performance. This could be performed by demonstrating whether there is a direct correlation between the degree of implementation of corporate governance and the increase in financial performance of the companies through research study. If the conclusion shows that there is a direct correlation between the degree of implementation of corporate governance (Ig) and the performance of companies (CA turnover), then recommendations can be made to improve corporate governance so that the performance of companies can increase even more, taking into account not only the interests of shareholders but also the interests of other participants in a company's business.

## 3. Materials and Methods

### 3.1. Research Materials and Methods according to the Literature

Several research studies have been developed internationally to demonstrate the correlation between governance and performance, using specific research methods.

We identified several studies that constructed an index of corporate governance and tested its relationship with firm performance, presented in Table 1. Many studies constructed an index from the components of corporate governance and measured performance based on ROE, ROA, or Tobin's Q indicators.

**Table 1.** Studies with index of corporate governance.

| Period of Analysis | Number of Companies/Country | Performance Measured as | Index Structure | Conclusion | Study |
|---|---|---|---|---|---|
| 2009–2014 | 407/India | return on net worth<br>return on assets | board structure, ownership structure;<br>market for corporate control;<br>product market competition | Positive relation with performance | [60] |
| 2007–2009 | Croatia | Tobin's Q | transparency of the business;<br>relations with shareholders;<br>the board of directors;<br>the supervisory board;<br>internal audit and control; | Strong relation with performance | [61] |
| 2016–2020 | 221/Malaysia | ROE<br>EBIT | ratio of family directors on board equals to or exceed those of independent directors;<br>the appointment of family women directors;<br>family director as the board chairman<br>family CEO duality | Negative relation with performance | [62] |
| 1999–2008 | 2212/UK | ROA<br>ROE<br>ROIC<br>Tobin's Q | percentage of NEDs on the board;<br>majority of NEDs independent;<br>presence of remuneration committee; presence of audit committee;<br>presence of nomination committee;<br>at least half of audit committee members are independent;<br>independent NED chairing audit committee | Positive relation with performance | [63] |
| 2009–2016 | 53/India and 53/Gulf Corporate Council countries | ROE<br>Tobin's Q | board accountability index;<br>audit committee index;<br>transparency disclosure index | Board accountability and audit committee have an insignificant impact on firms' performances | [64] |
| 2010–2017 | 26/Brazil | Market value | composition of the board of directors<br>ownership and control structure<br>compensation to managers;<br>protection of minority shareholders<br>transparency; and quality of the independent audit | Positive relation with performance | [65] |
| 2010–2016 | 116/Brazil | | (i) ownership and control structure; (ii) information disclosure and transparency; (iii) composition of the board of directors; (iv) management incentives; (v) shareholder rights | Composition of the board of director was the one with the greatest weight in the composition of the governance index. | [66] |
| 2009–2016 | 84/Jordan | Market share | disclosure and transparency;<br>board effectiveness and composition;<br>shareholders' rights | Firms with better overall corporate governance have better performance | [67] |

### 3.2. Sample Selection and Data Collection

In order to fulfill the research objectives of the present study, several steps were taken. Thus, the data of 43 energy companies in Romania were extracted, out of a total of 85, operating in the fields of electricity generation, transmission and distribution, natural gas, and oil. The sample is representative, and the data were extracted based on reports published annually in the period 2012–2021. We note that data were not yet available for all selected companies at the time the case study was developed for the financial year 2022. The data extracted for the companies that were the subject of the study related to turnover (CA), net profit (Pn), and number of employees (Ns) recorded at the end of the financial year.

### 3.3. Presentation of the Validation Model

We proposed the following hypotheses:

**H1.** *There is a significant direct correlation between the governance of energy companies and their performance.*

**H2.** *The governance of energy companies has a significant influence on the variation in energy company performance.*

The data obtained were centralized for each company, and then the correlation between performance and governance was analyzed using a simple linear regression model presented below:

$$y = x\,\alpha + \varepsilon, \tag{1}$$

where

- y is the dependent variable (explained, endogenous, resultant), in the case study represented by the performance (Ca) of the companies analyzed;
- x is the vector of independent variables (explanatory, exogenous), of dimension $1 \times p$, in the case study represented by the governance index (Ig) obtained using the scoring method;
- $\alpha$ is the vector of coefficients, of dimension $p \times 1$, the parameters of the model;
- $\varepsilon$ is a variable, interpreted as an error (disturbance, measurement error).

For these companies, data on the corporate governance system were also extracted, and a governance index (Ig) was determined using the scoring method. Thus, the maximum number of points for each company was 10, and the minimum number was 0 points (for companies that did not implement a corporate governance system).

In determining the scoring score, information such as general principles, risk management system, and internal control, fair rewards, motivation, and adding value through investor relations were considered. Each system of questions received a maximum of 2.5 points.

Information on general principles refers to the implementation of a corporate governance code, the existence of transparency regarding the members of the governing bodies, and the existence of specialized committees such as audit, nomination, and remuneration committees.

Information on the internal management and control system refers to the organization by companies of internal audit activity for the purpose of independently assessing, on a regular basis, the safety and effectiveness of the risk management and internal control system and corporate governance practices.

Information on fair reward and motivation refers to the existence of transparency regarding the remuneration policies of members of management bodies.

Information on adding value through investor relations refers to the existence of transparency regarding the organization of general meetings of shareholders/members and the respect of their voting rights.

In order to measure the degree of implementation of corporate governance in the companies covered by the research study, we used the scoring method, calculating a score for each company based on a set of questions presented in Table 2.

**Table 2.** Estimating the Governance Index (Ig) of Romanian energy companies.

| Test Number | Description of the Question | Score |
|---|---|---|
| T1 | Does the company publish a governance code? | If the answer is "YES", score = 0.25, otherwise the score is 0 |
| T2 | Does the company publish policies to combat conflicts of interest? | If the answer is "YES", score = 0.25, otherwise the score is 0 |
| T3 | Does the Management Board consist of at least five members? | If the answer is "YES", score = 0.25, otherwise the score is 0 |
| T4 | Are there independent board members? | If the answer is "YES", score = 0.25, otherwise the score is 0 |
| T5 | Does the company disclose information about any relationship with a shareholder who directly or indirectly owns shares representing more than 5% of all voting rights? | If the answer is "YES", score = 0.25, otherwise the score is 0 |
| T6 | Does the company publish information on the existence of a secretary on the Board responsible for supporting the work of the Board? | If the answer is "YES", score = 0.25, otherwise the score is 0 |
| T7 | Is there information on the evaluation of members of the Management Board? | If the answer is "YES", score = 0.25, otherwise the score is 0 |
| T8 | Does the company publish information about the number of board meetings? | If the answer is "YES", score = 0.25, otherwise the score is 0 |
| T9 | Does the corporate governance statement include information on the exact number of independent members? | If the answer is "YES", score = 0.25, otherwise the score is 0 |
| T10 | Does the company publish information on the existence of Nomination Committees? | If the answer is "YES", score = 0.25, otherwise the score is 0 |
| T11 | Does the company publish information on the existence of Audit Committees? | If the answer is "YES", score = 0.25, otherwise the score is 0 |
| T12 | Is the chairman of the Audit Committee an independent non-executive member? | If the answer is "YES", score = 0.25, otherwise the score is 0 |
| T13 | Does the Audit Committee review the internal control system annually? | If the answer is "YES", score = 0.25, otherwise the score is 0 |
| T14 | Does the company publish internal audit information? | If the answer is "YES", score = 0.25, otherwise the score is 0 |
| T15 | Does the Audit Committee assess conflicts of interest in relation to transactions of the company and its subsidiaries with related parties? | If the answer is "YES", score = 0.25, otherwise the score is 0 |
| T16 | Does the Audit Committee assess the effectiveness of the internal control system and the risk management system? | If the answer is "YES", score = 0.25, otherwise the score is 0 |
| T17 | Does the Audit Committee monitor the application of internal audit standards? | If the answer is "YES", score = 0.25, otherwise the score is 0 |
| T18 | Does the company publish information on audit committee reports to management? | If the answer is "YES", score = 0.25, otherwise the score is 0 |
| T19 | Does the company publish policies on equal treatment of shareholders? | If the answer is "YES", score = 0.25, otherwise the score is 0 |
| T20 | Does the company publish policies on transactions with shareholders holding more than 5% of the share capital? | If the answer is "YES", score = 0.25, otherwise the score is 0 |
| T21 | Does the company publish information on the outsourcing of internal audit? | If the answer is "YES", score = 0.25, otherwise the score is 0 |
| T22 | Does the Internal Audit Department report functionally to the Board through the Audit Committee? | If the answer is "YES", score = 0.25, otherwise the score is 0 |

**Table 2.** *Cont.*

| Test Number | Description of the Question | Score |
|---|---|---|
| T23 | Does the company publish information on the remuneration policy for members of the governing bodies? | If the answer is "YES", score = 0.25, otherwise the score is 0 |
| T24 | Does the company publish an investor relations section on its website? | If the answer is "YES", score = 0.25, otherwise the score is 0 |
| T25 | Does the company publish its updated Articles of Association on its website? | If the answer is "YES", score = 0.25, otherwise the score is 0 |
| T26 | Are the professional CVs of the members of the governing bodies published on the website? | If the answer is "YES", score = 0.25, otherwise the score is 0 |
| T27 | Are current financial reports and regular reports (quarterly, half yearly, and annual) published on the website? | If the answer is "YES", score = 0.25, otherwise the score is 0 |
| T28 | Does the company publish information on general meetings of shareholders on its website? | If the answer is "YES", score = 0.25, otherwise the score is 0 |
| T29 | Does the company publish information on dividend payments? | If the answer is "YES", score = 0.25, otherwise the score is 0 |
| T30 | Does the company publish on the website the contact details of a person to provide information on request? | If the answer is "YES", score = 0.25, otherwise the score is 0 |
| T31 | Does the company publish information on the statutory auditor's reports? | If the answer is "YES", score = 0.25, otherwise the score is 0 |
| T32 | Does the company publish information on its dividend distribution policy? | If the answer is "YES", score = 0.25, otherwise the score is 0 |
| T33 | Does the company publish information on revenue and expenditure forecasts? | If the answer is "YES", score = 0.25, otherwise the score is 0 |
| T34 | Does the company's management respect the right of shareholders to attend general meetings? | If the answer is "YES", score = 0.25, otherwise the score is 0 |
| T35 | Do the external auditors attend the general meeting of shareholders? | If the answer is "YES", score = 0.25, otherwise the score is 0 |
| T36 | Does the Board of Directors provide shareholders with a brief assessment of the internal control and significant risk management systems? | If the answer is "YES", score = 0.25, otherwise the score is 0 |
| T37 | Can journalists attend general meetings of shareholders? | If the answer is "YES", score = 0.25, otherwise the score is 0 |
| T38 | Do the quarterly and half yearly financial reports include information in both Romanian and English? | If the answer is "YES", score = 0.25, otherwise the score is 0 |
| T39 | Does the company hold press conferences to present its financial results? | If the answer is "YES", score = 0.25, otherwise the score is 0 |
| T40 | Does the company publish information on social responsibility? | If the answer is "YES", score = 0.25, otherwise the score is 0 |

## 4. Results

To estimate the coefficients of a linear model by the least squares method and to calculate the necessary statistics for the associated statistical tests, the Regression procedure, one of the most complex in the Excel statistical processing package, was used.

Before using this model, we centralized the companies' data, i.e., their governance index (Ig) and their performance, represented by their turnover as of 31 December 2021, as shown below in Table 3.

Next, we processed the data obtained using the Regression procedure from the Excel statistical processing package and using EViews software. After processing the data, we obtained the following results (Table 4).

**Table 3.** Governance and performance situation of Romanian energy companies.

| Company | Governance (Ig) | Performance (CA) -Mil. Euro- |
|---|---|---|
| C1 | 0 | 1.83 |
| C2 | 0 | 4.82 |
| C4 | 0 | 13.12 |
| C5 | 0 | 126.67 |
| C7 | 0 | 4.48 |
| C8 | 0 | 42.34 |
| C9 | 0 | 2.37 |
| C10 | 0 | 0.26 |
| C12 | 0 | 13.65 |
| C15 | 0 | 60.29 |
| C18 | 0 | 215.19 |
| C20 | 0 | 7.99 |
| C21 | 0 | 1894.11 |
| C22 | 0 | 12.54 |
| C24 | 0 | 3.83 |
| C25 | 0 | 39.13 |
| C31 | 0 | 8.88 |
| C33 | 0 | 2.20 |
| C27 | 1 | 1.16 |
| C32 | 2 | 6.18 |
| C14 | 2.5 | 2.58 |
| C28 | 3 | 4.85 |
| C41 | 3 | 216.29 |
| C26 | 4 | 721.15 |
| C29 | 4 | 95.95 |
| C42 | 4 | 766.36 |
| C36 | 5 | 1378.70 |
| C11 | 6 | 1201.22 |
| C19 | 6 | 1571.10 |
| C43 | 6 | 1599.26 |
| C6 | 8 | 489.51 |
| C16 | 8 | 654.97 |
| C17 | 8 | 677.02 |
| C3 | 10.00 | 666.53 |
| C13 | 10 | 68.86 |
| C23 | 10 | 934.52 |
| C30 | 10 | 82.98 |
| C34 | 10 | 662.50 |
| C35 | 10 | 1280.37 |
| C37 | 10 | 11.96 |

**Table 3.** *Cont.*

| Company | Governance (Ig) | Performance (CA) -Mil. Euro- |
|---|---|---|
| C38 | 10 | 1157.05 |
| C39 | 10 | 4766.70 |
| C40 | 10 | 2455.41 |

Source: own processing based on data extracted from the website www.mfinante.gov.ro [68].

**Table 4.** Summary output.

| **Regression Statistics** | |
|---|---|
| Multiple R | 0.510 |
| R Square | 0.260 |
| Adjusted R Square | 0.242 |
| Standard Error | 789.119 |
| Observations | 43 |

The correlation ratio (Multiple R) has a value of 0.51, indicating the existence of a correlation of moderate intensity between the two variables. At the same time, the regression model explains 26% of the variation in company performance (i.e., turnover)—in other words, 26% of the total variation in turnover is explained by changes in corporate governance.

The regression model is valid at a 5% significance level (Significance F < 0.05) (Table 5).

**Table 5.** ANOVA.

| | df | SS | MS | F | Significance F |
|---|---|---|---|---|---|
| Regression | 1 | 8,969,718.298 | 8,969,718.298 | 14.40436752 | 0.000478229 |
| Residual | 41 | 25,531,037.7 | 622,708.2366 | | |
| Total | 42 | 34,500,756 | | | |

The slope parameter of the regression model (the coefficient of the explanatory variable Ig) is statistically significant for a significance level of 5%, showing that we can expect the company's performance (turnover) to increase on average by EUR 111.13 million when improving the Governance Index by one scoring point (Table 6). Residual analysis with prediction error values and standardized error values is presented in Table 7. The probabilities are presented in Table 8. The residual diagram (Figure 2) and Line Fit Plots (Figure 3) allow analysis of the quality of the chosen model.

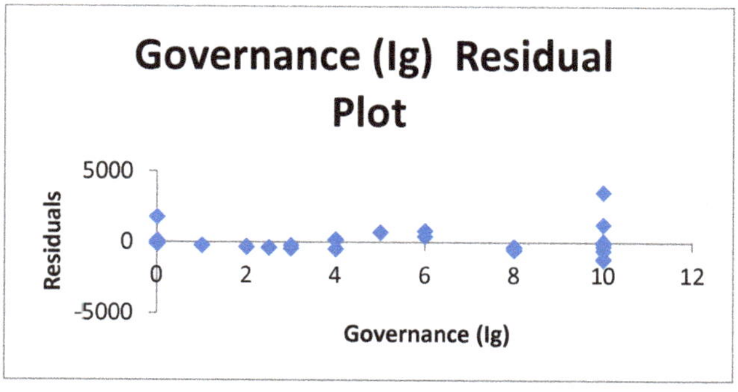


**Figure 2.** Residual Plot.

**Table 6.** Determining the Slope Parameter of the Regression Model.

| | Coefficients | Standard Error | t Stat | *p*-Value | Lower 95% | Upper 95% | Lower 95.0% | Upper 95.0% |
|---|---|---|---|---|---|---|---|---|
| Intercept | 115.7653 | 167.2220 | 0.6923 | 0.49266 | −221.9464 | 453.4769 | −221.9464 | 453.4769 |
| Governance (Ig) | 111.1378 | 29.2830 | 3.7953 | 0.00048 | 51.9997 | 170.2760 | 51.9997 | 170.2760 |

**Table 7.** Residual Output.

| Observation | Predicted Performance (CA) -Mil. Euro- | Residuals | Standard Residuals |
|---|---|---|---|
| 1 | 115.765261 | −113.93289 | −0.146130053 |
| 2 | 115.765261 | −110.9436 | −0.142295993 |
| 3 | 115.765261 | −102.64588 | −0.131653356 |
| 4 | 115.765261 | 10.909086 | 0.013991967 |
| 5 | 115.765261 | −111.28643 | −0.142735702 |
| 6 | 115.765261 | −73.42977 | −0.094180844 |
| 7 | 115.765261 | −113.39561 | −0.145440933 |
| 8 | 115.765261 | −115.50222 | −0.148142864 |
| 9 | 115.765261 | −102.11388 | −0.130971013 |
| 10 | 115.765261 | −55.470699 | −0.071146584 |
| 11 | 115.765261 | 99.429208 | 0.12752766 |
| 12 | 115.765261 | −107.77722 | −0.138234805 |
| 13 | 115.765261 | 1778.3479 | 2.280904729 |
| 14 | 115.765261 | −103.22537 | −0.132396611 |
| 15 | 115.765261 | −111.93315 | −0.143565187 |
| 16 | 115.765261 | −76.634376 | −0.098291064 |
| 17 | 115.765261 | −106.88319 | −0.137088122 |
| 18 | 115.765261 | −113.56748 | −0.145661371 |
| 19 | 226.9031069 | −225.74008 | −0.289533674 |
| 20 | 338.0409528 | −331.85871 | −0.425641177 |
| 21 | 393.6098758 | −391.03336 | −0.50153844 |
| 22 | 449.1787987 | −444.32406 | −0.569888961 |
| 23 | 449.1787987 | −232.88435 | −0.298696905 |
| 24 | 560.3166447 | 160.83067 | 0.20628103 |
| 25 | 560.3166447 | −464.36883 | −0.595598329 |
| 26 | 560.3166447 | 206.047 | 0.264275377 |
| 27 | 671.4544906 | 707.24797 | 0.907114523 |
| 28 | 782.5923365 | 418.62893 | 0.536932441 |
| 29 | 782.5923365 | 788.50678 | 1.011336882 |
| 30 | 782.5923365 | 816.66742 | 1.047455651 |
| 31 | 1004.868028 | −515.35509 | −0.660993178 |
| 32 | 1004.868028 | −349.90275 | −0.44878442 |
| 33 | 1004.868028 | −327.84737 | −0.420496244 |
| 34 | 1227.14372 | −560.61553 | −0.71904411 |
| 35 | 1227.14372 | −1158.2797 | −1.485606768 |
| 36 | 1227.14372 | −292.62162 | −0.375315778 |
| 37 | 1227.14372 | −1144.1595 | −1.467496224 |
| 38 | 1227.14372 | −564.64827 | −0.724216499 |
| 39 | 1227.14372 | 53.224597 | 0.068265739 |
| 40 | 1227.14372 | −1215.1867 | −1.558595541 |
| 41 | 1227.14372 | −70.090776 | −0.089898259 |
| 42 | 1227.14372 | 3539.552 | 4.539820825 |
| 43 | 1227.14372 | 1228.2669 | 1.575372183 |

**Table 8.** Probability output.

| Percentile | Performance (CA) -Mil. Euro- |
|---:|---:|
| 1.1627907 | 0.263043 |
| 3.4883721 | 1.16303 |
| 5.8139535 | 1.832367 |
| 8.1395349 | 2.197783 |
| 10.465116 | 2.369651 |
| 12.790698 | 2.576513 |
| 15.116279 | 3.83211 |
| 17.44186 | 4.478833 |
| 19.767442 | 4.82166 |
| 22.093023 | 4.854736 |
| 24.418605 | 6.182241 |
| 26.744186 | 7.988037 |
| 29.069767 | 8.882069 |
| 31.395349 | 11.95699 |
| 33.72093 | 12.53989 |
| 36.046512 | 13.11938 |
| 38.372093 | 13.65138 |
| 40.697674 | 39.13089 |
| 43.023256 | 42.33549 |
| 45.348837 | 60.29456 |
| 47.674419 | 68.86399 |
| 50 | 82.98419 |
| 52.325581 | 95.94782 |
| 54.651163 | 126.6743 |
| 56.976744 | 215.1945 |
| 59.302326 | 216.2944 |
| 61.627907 | 489.5129 |
| 63.953488 | 654.9653 |
| 66.27907 | 662.4954 |
| 68.604651 | 666.5282 |
| 70.930233 | 677.0207 |
| 73.255814 | 721.1473 |
| 75.581395 | 766.3636 |
| 77.906977 | 934.5221 |
| 80.232558 | 1157.053 |
| 82.55814 | 1201.221 |
| 84.883721 | 1280.368 |
| 87.209302 | 1378.702 |
| 89.534884 | 1571.099 |
| 91.860465 | 1599.26 |
| 94.186047 | 1894.113 |
| 96.511628 | 2455.411 |
| 98.837209 | 4766.696 |

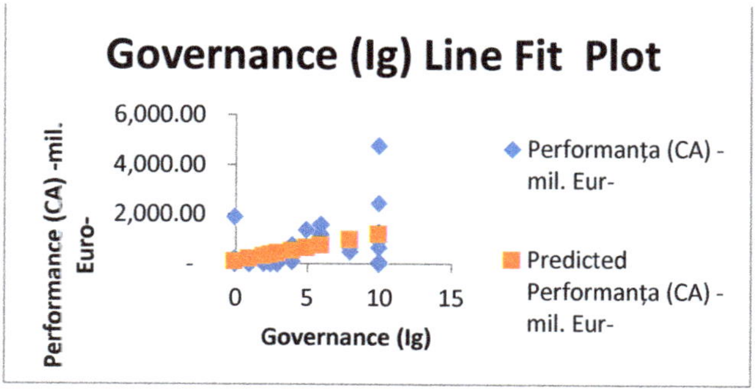


**Figure 3.** Line Fit Plot.

The linear regression equation is:

$$y = 115.77 + 111.14 \cdot x \quad (2)$$
$$(167.22) \quad (29.28)$$

The correlation matrix (Table 9) indicates that there is a direct correlation of moderate strength between the two variables, and the Pearson correlation coefficient (of 0.51) is statistically significant at a 5% significance level. Hypothesis H1 (There is a significant direct correlation between the governance of energy companies and their performance) is validated by identifying a Pearson correlation coefficient with a positive, statistically significant value.

**Table 9.** Correlation Matrix.

| | | Performance_CA | Governance_Ig |
|---|---|---|---|
| Pearson Correlation | Performance_CA | 1.000 | 0.510 |
| | Governance_Ig | 0.510 | 1.000 |
| Sig. (1-tailed) | Performance_CA | | 0.000 |
| | Governance_Ig | 0.000 | |
| N | Performance_CA | 43 | 43 |
| | Governance_Ig | 43 | 43 |

The correlogram (Figure 4) indicates the existence of a direct link between the variables, and the linear regression equation modelling the trend in this link is presented, expressing the dependence of company performance on corporate governance.

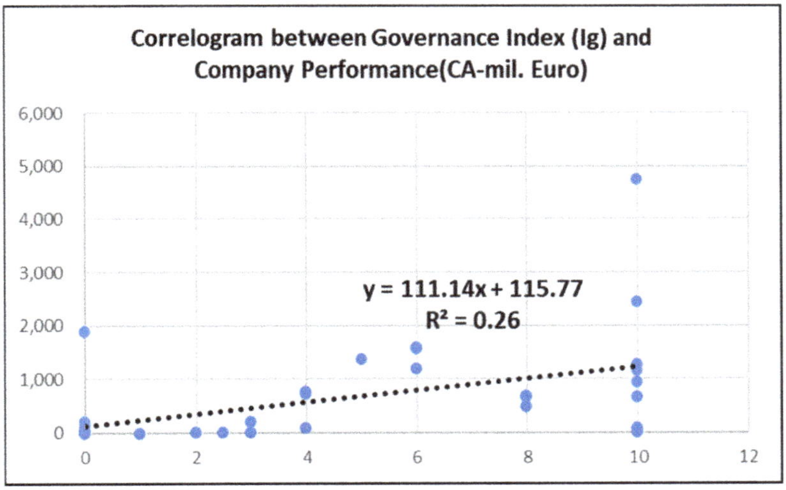


**Figure 4.** Correlogram between Governance Index and Performance.

In order to refine the model presented above to ensure the validation of all assumptions of the linear regression model and to improve its ability to explain the variation in the response variable, this variable was log transformed and included in a shape model:

$$\ln(CA_i) = \beta_0 + \beta_1 \cdot Ig_i + \varepsilon_i,\ i = \overline{1, N} \quad (3)$$

where

β_0, β_1 are the model parameters (Intercept and Slope parameters);
i = energy company;

$CA_i$ = performance quantified by the turnover of energy company i;
$Ig_i$ = Governance index of energy company i;
N = population volume.

The application of the natural logarithm function to the values of the response variable can be beneficial for model optimization if outliers are observed between the values of this variable.

Applying the log–linear regression method, the following model resulted in:

$$\ln(CA_i) = 2.618 + 0.419 \cdot Ig_i \quad (4)$$

The model is statistically valid at 5% significance level (Sig = 0.0000015 < 0.05) (Table 10), both parameters (Intercept and Slope) are statistically significant at the same significance level (Sig(β_0) = 0.0000003 < 0.05; Sig(β_1) = 0.0000015 < 0. 05) (Table 11), with the sign of the slope parameter indicating a direct correlation between the governance of energy companies and their performance, with an increase in the governance index leading to a relative increase in turnover.

**Table 10.** ANOVA [a].

| Model | | Sum of Squares | df | Mean Square | F | Sig. |
|---|---|---|---|---|---|---|
| 1 | Regression | 127.248 | 1 | 127.248 | 31.682 | 0.0000015 [b] |
| | Residual | 164.673 | 41 | 4.016 | | |
| | Total | 291.921 | 42 | | | |

[a]. Dependent Variable: LN_CA. [b]. Predictors: (Constant), Ig.

**Table 11.** Coefficients [a].

| Model | | Unstandardized Coefficients | | Standardized Coefficients | t | Sig. |
|---|---|---|---|---|---|---|
| | | B | Std. Error | Beta | | |
| 1 | (Constant) | 2.618 | 0.425 | | 6.165 | 0.0000003 |
| | Ig | 0.419 | 0.074 | 0.660 | 5.629 | 0.0000015 |

[a]. Dependent Variable: LN_CA.

In addition, the model has a higher explanatory power than the previous one, explaining 43.6% of the relative variation in company performance (Table 12).

**Table 12.** Model Summary [b].

| Model | R | R Square | Adjusted R Square | Std. Error of the Estimate | Durbin-Watson |
|---|---|---|---|---|---|
| 1 | 0.660 [a] | 0.436 | 0.422 | 2.00410 | 1.417 |

[a]. Predictors: (Constant), Ig. [b]. Dependent Variable: LN_CA.

The quality of the model was also assessed through the analysis of the distribution of residuals (testing the hypothesis of normality of errors, the hypothesis of homoscedasticity of errors and the hypothesis of non-homoscedasticity of errors). Verification of the normality of the distribution of the residuals was checked both by graphical procedures (Error Histogram and Normal P-P Plot) and by numerical procedures: Kolmogorov–Smirnov and Shapiro–Wilk tests. (Figures 5 and 6).

The results demonstrated the validation of the hypothesis of normality of the errors, at a significance level of 5% (Sig.(Shapiro–Wilk) = 0.792 > 0.05, accepting the null hypothesis of the test, according to which there is no significant difference between the distribution of errors and the normal distribution) (Table 13).

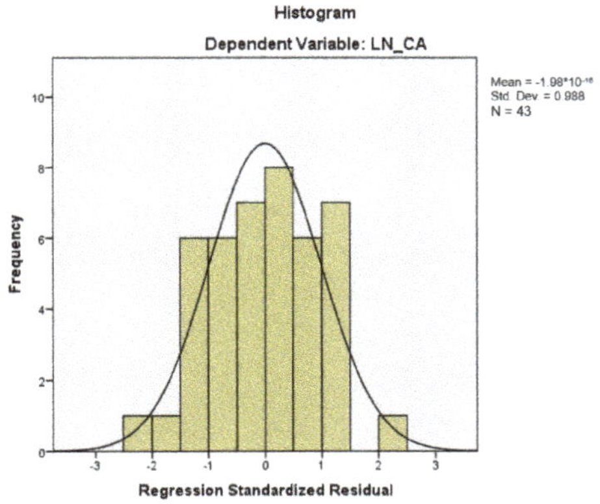


**Figure 5.** Histogram.

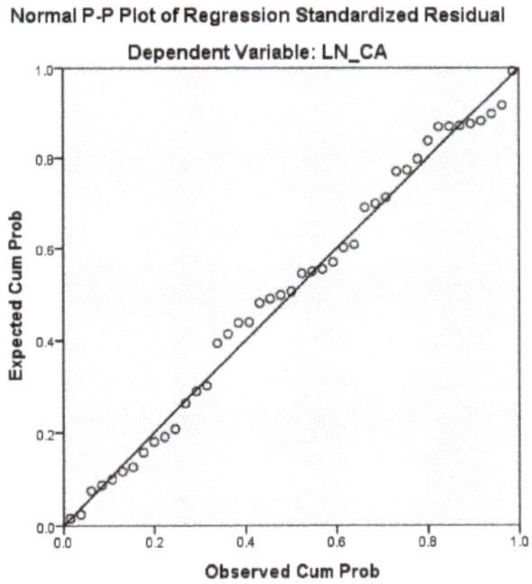


**Figure 6.** Normal P-P Plot.

**Table 13.** Tests of Normality.

| | Kolmogorov–Smirnov [a] | | | Shapiro–Wilk | | |
|---|---|---|---|---|---|---|
| | **Statistic** | **df** | **Sig.** | **Statistic** | **df** | **Sig.** |
| Standardized Residual | 0.067 | 43 | 0.200 * | 0.984 | 43 | 0.792 |

* This is a lower bound of the true significance. [a]. Lilliefors Significance Correction.

The hypothesis of homoscedasticity of errors was tested using the White Test, yielding the following auxiliary regression model:

$$e^2 = 4.226 - 0.014 \cdot IG - 0.01 \cdot Ig^2 \tag{5}$$

The calculated value of the White (Lagrange Multiplier) test (0.39) is lower than the critical value (Hi_patrat_crit = 5.99), leading to acceptance of the null hypothesis of the test, which supports homoscedasticity of errors (5% significance level, $p$ value = 0.82 > 0.05) (Table 14).

The hypothesis of non-autocorrelation of errors was tested using the Durbin Watson test, whose calculated value obtained is 1.417 (see Table 12) (critical values at 5% significance level: 1.246 and 1.344). The conclusion following the application of the test is to accept the null hypothesis that the errors are not autocorrelated.

**Table 14.** Auxiliary model results.

| Auxiliary R Square | 0.009184 | Coefficients | |
|---|---|---|---|
| Lagrange Multiplier | 0.394910 | Intercept | 4.226 |
| Chi_Square_crit | 5.991465 | Ig | −0.014 |
| *p* Value | 0.820817 | $Ig^2$ | −0.010 |

Following the validation of the regression model and its quality assessment, Working Hypothesis 2 is verified, i.e., the governance of energy companies has a significant influence on the variation in energy company performance.

## 5. Discussion

The results of the study show that the working hypotheses are verified following the validation of the regression model and the assessment of its quality; thus, the governance of energy companies has a significant influence on the variation in the performance of energy companies in Romania. Thus, it is shown that the higher the corporate governance (Ig) system the higher the performance of energy companies. More specifically, energy companies that have implemented governance rules have a higher turnover than companies that have not implemented a governance system or where the corporate governance index is low.

From the analysis of the results of the study, it can be noted that, in general, companies listed on a regulated market (Bucharest Stock Exchange) obtained a maximum score of the governance index (Ig), i.e., 10 points. It is worth noting that their turnover is also among the highest of the 43 companies surveyed. Thus, two companies in the oil sector that are listed on the stock exchange stand out, with a turnover of EUR 4766 million and EUR 2455 million, respectively, in the 2021 financial year, a listed company in the field of electricity production by hydropower plants, which achieved a turnover of EUR 1280 million in the 2021 financial year, and a listed company in the field of natural gas extraction, which achieved a turnover of EUR 1157 million, and a company producing coal-based electricity, which achieved a turnover of EUR 662 million in the 2021 financial year. From the analysis of the survey results, we found that those companies for which we estimated a low governance index also have poor financial performances. Thus, the lowest turnovers were obtained by some companies that buy and sell electricity, where the governance index (Ig) is zero, and the turnover is 263,000 euros, i.e., 1.8 million euros.

The results of the study show the importance of the energy sector in a country's economy, by contributing to the formation of gross domestic product, the formation of budget revenues, and the provision of jobs.

According to the study, the cumulative turnover of the 43 companies surveyed for the period 2012–2021 is approximately EUR 154.68 billion, of which EUR 23.9 billion is the turnover achieved in 2021. Analyzing the turnover dynamics, in the financial year 2021, there was an increase in turnover compared to the financial year 2020 for most of the energy companies surveyed, especially among the large electricity producers that recorded turnover increases of even more than 60%. This increase in turnover was due to an increase in sales prices to distributors and consumers as well as an increase in the quantity of electricity sold. In terms of natural gas production and sales, seven companies had an increase in turnover in 2021 compared to 2020 between 18% and 84%. In the period 2012–2021, the surveyed companies achieved a cumulative profit of about EUR 7.95 billion, of which EUR 1.54 billion was the cumulative profit for the surveyed companies achieved in the financial year 2021. The average cumulative turnover profitability for the 43 companies for the period 2012–2021 was 5.14%. In the financial year 2021, the average return on aggregate turnover for the 43 companies was 6.45%, and it is worth noting the return of even 47.66% obtained by one of the largest electricity producers in Romania, which has a majority state capital, which is quite attractive in the capital market.

The energy companies under study play a very important role in the macroeconomic stability of the country, providing jobs for 52,919 employees, according to the financial reports as of 31 December 2021.

After analyzing the results of the study, we found that there is a direct correlation between the governance and financial performance of energy companies. Thus, the higher the governance index the higher the financial performance of the companies. This conclusion is justified by the following aspects:

- greater transparency in the appointment of board members leads to the selection of the best professionally trained people so that their decisions will also lead to increased financial performance and shareholder value of companies;
- implementation of corporate governance principles regarding the separation of the control function from the executive function leads to a streamlining of the work of the members of the Management Board, the Supervisory Board, the Executive Board, and the Executive Management;
- the functioning of the Nomination Committee and the Remuneration Committee contributes to increasing transparency regarding the appointment and remuneration of the members of the governing bodies, the basic principle being that the payment of directors' fees should be in line with the financial performance of the companies;
- the establishment of specialized committees, such as the audit committee, helps to streamline corporate decision-making by facilitating communication between those responsible for corporate governance and the external auditor;
- implementing a risk management system makes companies more efficient when control is strengthened by appointing internal auditors;
- the appointment of statutory auditors, together with increased transparency on financial reporting, increases the confidence of current and potential investors in the financial results reported by companies, helping to attract new capital;
- greater transparency of company decisions leads to greater respect for shareholders' rights, such as the right to be informed, the right to vote, the right to be elected, the right to receive dividends.

All the corporate governance principles mentioned above contribute to making companies more efficient, maximizing profits, and maximizing dividends for shareholders.

We believe that the results of the study may be important for the following reasons: firstly, the national, European, and even international energy market is in a difficult period due to the regional geopolitical situation in Ukraine. Thus, even before the beginning of the conflict, the price of gas and electricity rose, which subsequently increased in Romania, leading to a sharp rise in inflation. Another reason why this study may be important is that the sample selected includes the most important energy companies. Thus, at the political level, there are discussions on the sale of shares in the most important hydropower producer, in which the Romanian state holds most shares. The most important companies in the fields of oil extraction and processing, gas extraction, transmission and distribution, and electricity distribution were also included in the study.

## 6. Conclusions

The implementation of corporate governance is optional in Romania. Our results could have positive implications for the business environment, as increased performance through the application of corporate governance is an incentive for the implementation of corporate governance. Corporate governance is a key element in improving economic efficiency and growth as well as increasing investor confidence. It provides the framework/context through which the company's objectives are set and the means to achieve these objectives.

The limitation of the study undertaken is the unavailability of the necessary data for some companies.

The estimated governance index of these companies (Ig) can be used in future studies to determine strengths and weaknesses and to improve the governance systems of the companies covered in this study.

Further research may consider other factors in the construction of the corporate governance index, such as the presence of women on the board of directors, age of directors, directors' education, etc.

**Author Contributions:** Conceptualization, A.D.B. and C.E.B.; methodology, A.D.B. and C.E.B.; software, A.D.B. and C.E.B.; validation, A.D.B. and C.E.B.; data analysis, A.D.B. and C.E.B.; writing—original draft preparation, A.D.B.; writing—review and editing, C.E.B.; funding acquisition, C.E.B. All authors have read and agreed to the published version of the manuscript.

**Funding:** The A.P.C was funded by the University Politehnica of Bucharest through its PubArt program.

**Data Availability Statement:** The data presented in this study are available on request from the corresponding author.

**Conflicts of Interest:** The authors declare no conflict of interest. The funders had no role in the design of the study; in the collection, analyses, or interpretation of data; in the writing of the manuscript; or in the decision to publish the results.

## References

1. OECD Principles of Corporate Governance. 1999. Available online: www.oecd.org (accessed on 15 March 2023).
2. International Standards on Auditing and Code of Ethics. International Federation of Accountants. 2009. Available online: www.ifac.org (accessed on 15 March 2023).
3. Bigioi, A.D. *Teoria Generala a Guvernantei Corporative*; Editura ASE: Bucharest, Romania, 2015.
4. Bigioi, A.D. Transparenta Raportarii Financiare, Normele si Profesia Contabila—In Sprijinul Guvernantei Corporative. Ph.D. Thesis, The Bucharest University of Economic Studies, Bucharest, Romania, 2012.
5. Feleaga, L.; Feleaga, N. Contabilitate Financiara: O Abordare Europeana si Internationala; Editura Economica: Bucharest, Romania, 200; Volume 1, pp. 149–168.
6. BNR Forcasts. Available online: https://www.bnr.ro/Proiectii-BNR-4351-Mobile.aspx (accessed on 27 April 2023).
7. Electricity & Gas Hit Record Prices in 2022. Available online: https://ec.europa.eu/eurostat/web/products-eurostat-news/w/DDN-20230426-2 (accessed on 27 April 2023).
8. Kubiczek, J.; Derej, W. Financial performance of businesses in the COVID-19 pandemic conditions—Comparative study. *Polish J. Manag. Stud.* **2021**, *24*, 183–201. [CrossRef]
9. Jebran, K.; Chen, S.H. Can we learn lessons from the past? COVID-19 crisis and corporate governance responses. *Int. J. Financ. Econ.* **2023**, *28*, 421–429. [CrossRef]
10. Khatib, S.F.A.; Nour, A. The impact of Corporate Governance on firm performance during the COVID-19 pandemic: Evidence from Malaysia. *J. Asian Financ. Econ. Bus.* **2021**, *8*, 943–952.
11. Hindasah, L.; Akmalia, A. Can Corporate Governance protect firm performance during the COVID-19 pandemic? *Qual. Access Success* **2023**, *24*, 174–182.
12. Dang, V.C.; Nguyen, Q.K. Internal corporate governance and stock price cash risk: Evidence from Vietnam. *J. Sustain. Financ. Investig.* **2021**, 1–18. [CrossRef]
13. Almustafa, H.; Nguyen, Q.K. The impact of COVID-19 on firm risk and performance in MENA countries: Does national governance quality matter? *PLoS ONE* **2023**, *18*, e0281148. [CrossRef]
14. Wieczorek-Kosmala, M.; Henschel, T. The role of ERM and corporate governance in managing COVID-19 impacts: SMEs perspective. *J. Risk Financ. Manag.* **2022**, *15*, 587. [CrossRef]
15. Rhode, D.; Packel, A.K. Diversity on corporate boards: How much difference does difference make? *Del. J. Corp. Law* **2014**, *39*, 377–426. [CrossRef]
16. Sarhan, A.A.; Ntim, C.G.; Al-Najjar, B. Board diversity, corporate governance, corporate performance, and executive pay. *Int. J. Financ. Econ.* **2019**, *24*, 761–786. [CrossRef]
17. Al-rahahleh, A.S. Corporate governance quality, board gender diversity and corporate dividend policy: Evidence from Jordan. *Australas. Account. Bus. Financ. J.* **2017**, *11*, 86–104. [CrossRef]
18. Dionne, G.; Triki, T. Risk management and corporate governance: The importance of independence and financial knowledge for the Board and the Audit Committee. *Risk Manag. Insur. Rev.* **2019**, *22*, 247–277. [CrossRef]
19. Melon-Izco, A.; Ruiz-Cabestre, F.J.; Ruiz-olalla, C. Determinants of good governance practices: The role of board independence. *Span. J. Financ. Account.-Rev. Esp. Financ. Contab.* **2020**, *49*, 370–393. [CrossRef]
20. Karim, S.; Manab, N.A. Assessing the Governance Mechanism, Corporate Social Responsibility and Performance: The moderating effect of Board Independence. *Glob. Bus. Rev.* **2020**, 0972150920917773. [CrossRef]
21. Dos Santos, J.M.P. Financial knowledge for corporate boards: What should board members know? *Indep. J. Manag. Prod.* **2021**, *12*, 265–281. [CrossRef]
22. Filipovic, M. Research on the relationship between governance structures and external audit in Corporate Governance. *Ekon. Pregl.* **2021**, *72*, 522–549.

23. Serra, S.; Lemos, K. The influence of corporate governance and audit in risk discosure. *Rev. Evid. Contab. Financ.* **2020**, *8*, 106–124. [CrossRef]
24. Cho, C.C.; Wu, C.H. Role of auditor in agency conflict and corporate governance. Empirical analyses of Taiwanese firms. *Chin. Manag. Stud.* **2014**, *8*, 333–353. [CrossRef]
25. Chen, G.; Firth, M.; Gao, N.D.; Rui, O.M. Do ownership Structure and Governance Mechanism Have an Effect on Corporate Fraud in China's Listed Firms? 2005. Available online: https://papers.ssrn.com/sol3/papers.cfm?abstract_id=728945 (accessed on 10 April 2023).
26. Lappalainen, J.; Niskanen, M. Financial Performance of SMEs-Evidence on the Impact of Ownership Structure and Board Composition. 2009. Available online: https://papers.ssrn.com/sol3/papers.cfm?abstract_id=1500853 (accessed on 10 April 2023).
27. Castaner, X.; Goranova, M.; Hermes, N. Ownership and corporate governance across institutional contexts. *Corp. Gov. Int. Rev.* **2022**, *30*, 638–655. [CrossRef]
28. Batra, S.; Saini, M.; Yadav, M. Mapping the intellectual structure of corporate governance and ownership structure: A bibliometric analysis. *Int. J. Law Manag.* **2023**, *65*, 333–353. [CrossRef]
29. Lu, L.W. The moderating effect of corporate governance on the relationship between corporate sustainability performance and corporate financial performance. *Int. J. Discl. Gov.* **2021**, *18*, 193–206. [CrossRef]
30. Nguyen, N.P.A.; Dao, T.T.B. Liquidity, corporate governance performance: A meta-analysis. *Cogent Bus. Manag.* **2022**, *9*, 2137960. [CrossRef]
31. Neralla, N.G. Can corporate governance structure effect on corporate performance: An empirical investigation from Indian companies. *Int. J. Discl. Gov.* **2022**, *19*, 282–300. [CrossRef]
32. Hapsari, D.W.; Yadiati, W.; Suharman, H.; Rosdini, D. The mediating impact of value chain in the link between corporate governance and SOEs performance. *Australas. Account. Bus. Financ. J.* **2023**, *17*, 75–85. [CrossRef]
33. Hermanto, Y.B.; Lusy, L.; Widyastuti, M. How financial performance and state-owned enterprise (SOE) values are affected by Good Corporate Governance and intellectual capital perspective. *Economies* **2021**, *9*, 134. [CrossRef]
34. Abang'a, A.O.; Tauringana, V.; Wang'ombe, D.; Achiro, L.O. Corporate governance and financial performance of state-owned enterprises in Kenya. *Corp. Gov. Int. J. Bus. Soc.* **2022**, *22*, 798–820. [CrossRef]
35. Csedo, Z.; Magyari, J.; Zavarko, M. Dynamic Corporate Governance, Innovation, and Sustainability: Post-COVID Period. *Sustainability* **2022**, *14*, 3189. [CrossRef]
36. Alrazi, B.; Husin, N.M. Institutional Governance Framework for Determining Carbon-related Accounting Practices: An Exploratory Study of Electricity Generating Companies in Malaysia. In Proceedings of the IOP Conference Series: Earth and Environmental Science, Kuala Lumpur, Malaysia, 23–25 February 2016.
37. Milojevic, M.; Urbanski, M.; Terzic, I.; Prasolov, V. Impact of Non-Financial Factors on the Effectiveness of Audits in Energy Companies. *Energies* **2020**, *13*, 6212. [CrossRef]
38. Fakoya, M.B.; Nakeng, M.V. Board characteristics and sustainable energy performance of selected companies in South Africa. *Sustain. Prod. Consum.* **2019**, *18*, 190–199. [CrossRef]
39. Zhang, D.Y.; Zhang, Z.W.; Ji, Q.; Lucey, B.; Liu, J. Board characteristics, external governance and the use of renewable energy: International evidence. *J. Int. Financ. Mark. Inst. Money* **2021**, *7*, 101317. [CrossRef]
40. Alhawaj, A.; Buallay, A.; Abdallah, W. Sustainability reporting and energy sectorial performance: Developed and emerging economies. *Int. J. Energy Sect. Manag.* **2022**, *17*, 739–760. [CrossRef]
41. Liu, P.D.; Zhu, B.Y.; Yang, M.Y.; Chu, X. ESG and financial performance: A qualitative comparative analysis in China's new energy companies. *J. Clean. Prod.* **2022**, *379*, 134721. [CrossRef]
42. Hurduzeu, G.; Noja, G.G.; Cristea, M.; Dracea, R.M.; Filip, R.I. Revisiting the impact of ESG practices on firm financial performance in the energy sector: New empirical evidence. *Econ. Comput. Econ. Cybern. Stud. Res.* **2022**, *56*, 37–53.
43. Zheng, S.N.; He, C.H.; Hsu, S.C.; Sarkis, J.; Chen, J.H. Corporate environmental performance prediction in China: An empirical study of energy service companies. *J. Clean. Prod.* **2020**, *266*, 121395. [CrossRef]
44. Robaina, M.; Madaleno, M. The relationship between emissions reduction and financial performance: Are Portuguese companies in a sustainable development path? *Corp. Soc. Responsib. Environ. Manag.* **2020**, *27*, 1876. [CrossRef]
45. Szczepankiewicz, E.I.; Blazynska, J.; Zaleska, B.; Ullah, F.; Loopesko, W.E. Compliance with Corporate Governance Principles by Energy Companies Compared with All Companies Listed on the Warsaw Stock Exchange. *Energies* **2022**, *15*, 6481. [CrossRef]
46. Wahyuningrum, I.F.S.; Budihardjo, M.A. Relationship between Company Financial Performance, Characteristic and Environmental Disclosure of ASX Listed Companies. In Proceedings of the 3rd International Conference on Energy, Environmental and Information System (ICENIS 2018), Semarang, Indonesia, 14–15 August 2018; Volume 73, p. 10024.
47. Georgakopoulos, G.; Toudas, K.; Poutos, E.I.; Kounadeas, T.; Tsavalias, S. Capital Structure, Corporate Governance, Equity Ownership and Their Impact on Firms' -Profitability and Effectiveness in the Energy Sector. *Energies* **2022**, *15*, 3625. [CrossRef]
48. Aljaaidi, K.S. Economic value of energy and petrochemical companies and multiple directorships: Evidence from Saudi Arabia. *AD-Minist.* **2022**, *41*, 149–166. [CrossRef]
49. Gong, X.; Lin, A.L.; Chen, X.Q. CEO-CFO gender congruence and stock price crash risk in energy companies. *Econ. Anal. Policy* **2022**, *75*, 591–609. [CrossRef]
50. Ma, Z.; Shu, G.; Wang, Q.; Wang, L.F. Sustainable Governance and Green Innovation: A Perspective from Gender Diversity in China's Listed Companies. *Sustainability* **2022**, *14*, 6403. [CrossRef]

51. Bashindzhagyan, A.; Kharchilava, K.; Belyayeva, I. Influence of Corporate Governance on Financial Performance Of Russian State-Owned Companies. In Proceedings of the Global Challenges and Prospects of the Modern Economic Development, Samara, Russia, 6–8 December 2018.
52. Grabinska, B.; Kedzior, M.; Kedzior, D.; Grabinski, K. The Impact of Corporate Governance on the Capital Structure of Companies from the Energy Industry. The Case of Poland. *Energies* **2021**, *14*, 7412. [CrossRef]
53. Shaikh, I. Environmental, social and governance (ESG) practice and firm performance: An international evidence. *J. Bus. Econ. Manag.* **2022**, *23*, 218–237. [CrossRef]
54. Constantinescu, D.; Mititean, P. Association of ESG factors' disclosure with the value of European companies from energy industry. In Proceedings of the 8th International Scientific Conference IFRS: Global rules and local use-beyond the numbers, Prague, Czech Republic, 8–9 October 2020.
55. Malanski, L.K. Vertical integration and corporate governance over electricity sector companies'profitability. *Rev. Gest. E Secr. GESEC* **2022**, *13*, 1711–1729. [CrossRef]
56. Paun, D. Sustainability and Financial Performance of Companies in the Energy Sector in Romania. *Sustainability* **2017**, *9*, 1722. [CrossRef]
57. Matei, F.B.; Boboc, C.; Ghita, S. The relationship between corporate social responsibility and financial performance in Romanian companies. *Econ. Comput. Econ. Cybern. Stud. Res.* **2021**, *55*, 297–314.
58. Bigioi, A.D.; Bigioi, C.E. Study on corporate governance at Romanian banking institutions. *Manag. Strateg. J.* **2018**, *41*, 90–97.
59. Bigioi, A.D.; Bigioi, C.E. Harmonization of national rules with the new international recommendations on corporate governance. Case study from insurance company in Romania. *Manag. Strateg. J.* **2017**, *34*, 105–113.
60. Arora, A.; Bodhanwala, S. Relationship between corporate governance index and firm performance: Indian evidence. *Glob. Bus. Rev.* **2017**, *19*, 675–689. [CrossRef]
61. Korent, D.; Dundek, I.; Calopa, M.K. Corporate governance practices and firm performance measured by Croatian Corporate Governance Index (CCGI®). *Econ. Res. Ekon. Istraz.* **2014**, *27*, 221–231. [CrossRef]
62. Ling, L.; Jong, L.; Law, W.; Chieng, F. Family director board governance index: An analysis of family directors and firm performance in Malaysia. *J. Gen. Manag.* **2023**. [CrossRef]
63. Shaukat, A.; Trojanowski, G. Board governance and corporate performance. *J. Bus. Financ. Account.* **2018**, *45*, 184–208. [CrossRef]
64. Al-ahdal, W.M.; Alsamhi, M.H.; Tabash, M.I.; Farhan, N.H. The impact of corporate governcane on financial performance of Indian and GCC listed firms: An empirical investigation. *Res. Int. Bus. Financ.* **2020**, *51*, 101083. [CrossRef]
65. Dos Santos Sant'Ana, N.L.; Pires Sant'Ana, P.C. Corporate governance index and its relationship to market value. *Rev. Gest. Tecnol.* **2021**, *21*, 49–75.
66. Dos Santos, T.A.; de Souza, A.A.; Pessanha, G.R.G. Corporate governance index for companies in the Brasilian stock market. *Rev. Gest. Financ. Contab.* **2019**, *9*, 72–92.
67. Mansour, M.; Hashim, H.A.; Salleh, Z.; Al-ahdal, W.M.; Almaqtari, F.A.; Qamhan, M.A. Governance practices and corporate performance: Assessing the competence of principal-based guidelines. *Cogent Bus. Manag.* **2022**, *9*, 2105570. [CrossRef]
68. Information about Legal Persons. Romanian Ministry of Finance. Available online: www.mfinante.gov.ro (accessed on 10 April 2023).

*energies*

Article

# Determinants of the Efficiency of Electricity Generation in Latin America and Caribbean Countries Using a Cragg's Regression Model

**Sebastian Cuadros [1], Yeny E. Rodríguez [2] and Javier Contreras [3,*]**

1 Financial and Accounting Department, ICESI University, Cali 760031, Colombia; sebastian.cuadros@u.icesi.edu.co
2 School of Management, Universidad de los Andes, Bogotá 111711, Colombia; ye.rodriguez395@uniandes.edu.co
3 E.T.S. de Ingeniería Industrial, University of Castilla-La Mancha, 13071 Ciudad Real, Spain
* Correspondence: javier.contreras@uclm.es

**Abstract:** Measuring and understanding the evolution of the electricity generation efficiency sector is a relevant task for policymakers. This study focuses on assessing the determinants of electricity generation efficiency across 24 countries in Latin America and the Caribbean between 2000 and 2020. A hurdle model was employed to account for the censored nature of the dependent variable. The findings show a positive impact of opening up the electricity sector in fully efficient countries. Market liberalization and the involvement of privately owned power plants contribute to efficiency gains in inefficient countries. Additionally, the tariff protectionism level exhibits positive effects for completely efficient and inefficient countries. These results provide valuable insights, as less-efficient countries can glean strategies from their more efficient counterparts to enhance their electricity generation efficiency.

**Keywords:** efficiency; data envelopment analysis; hurdle model; openness; power generation

## 1. Introduction

Trade between different markets and different products has positive effects on productive efficiency. One explanation of this relationship is the complementarity of pro-duction and demand patterns between the markets that take part in the exchange, for ex-ample, substituting the most expensive generation with the cheapest energy from imports. The benefit of the exchange derives from the scale of trade as well as the geographic scope of the interconnected markets [1].

In electricity generation, the potential gains from regional interconnection are associated with complementarities in the demand for electricity and the different endowments of primary energy resources among trading countries due to differences in seasonal generation and consumption patterns [2,3]. Interconnections can provide important benefits such as (a) security of supply [4,5], (b) efficiency in generation costs [3,6] (c) increased competition [3], and (d) a decrease in environmental costs [3].

The risk of generation system failure is referred to as supply security. In addition, the stochastic demand behaviour generates uncertainties in the required quantity, leading to a higher demand than the supply on some occasions. Faced with this situation and considering the inelasticity of electricity demand in the short term, regional interconnection increases supply security since electricity imports allow full demand coverage [4,5].

Different studies mention that interconnections allow an optimal allocation of resources for electricity generation, unlike autarky between countries, which induces excessive levels of generation capacity and unnecessary and risky investments in new plants that translate into high production costs [3,6]. In addition, the excessive capacity levels resulting from autarky in the electricity sector increase the price for the final consumer [4] and the volatility in prices in the different regions [7]. According to Mulder and Giesbertz [7], interconnection allows more consumers to access the electricity service because the marginal cost would otherwise exceed some consumers' willingness to pay.

The increase in competition is evidenced in the integration between border markets, which promotes the entry of new companies with different sources of generation that minimize production costs and increase production efficiency. Finally, the decrease in environmental costs is caused by the increase in using new generation sources based on renewable resources such as wind and sun. In addition, the increase in the use of these clean sources reduces the challenges of balancing the energy system generated by their intermittent nature [3].

Despite the benefits of electricity interconnection, political and regulatory aspects largely determine its scope. Sometimes, the reforms create entry barriers for new competitors in regional markets. For example, price subsidies reduce incentives to expand investment in export-oriented electricity generation to neighbouring countries and limit the benefits of expanding transmission across borders [2].

Considering the foregoing, we aim to evaluate the determinants of electricity generation efficiency in 24 countries of Latin America and the Caribbean from 2000 to 2020, to identify relevant factors such as the openness of the electricity sector as a proxy of integration between countries. We use the model proposed by Cuadros et al. [8], who measured—from a dynamic DEA model using desirable and undesirable outputs—the efficiency of the sector. The novelty of this paper is focused not only on evaluating the factors that explain the electricity generation efficiency of Latin American countries, but also on identifying policy measures that can be carried on by these countries to improve their efficiency levels. From the methodology, we can distinguish between efficient and nonefficient countries in terms of the electricity generation activity. This helps us suggest policy measures that improve efficiency in inefficient countries by investing in the identified factors that impact their productivity, and establishing the factors have allowed efficient countries leverage their productive capacity.

Further material is divided into several parts. Thus, in Section 2, we present the major studies related to the determinants of electricity generation efficiency. In Section 3, we present the methodology used to estimate the efficiency model. In Section 4, we show the descriptive statistics of the variables used in the efficiency evaluation and present the model results. Finally, in Section 5, we present the conclusions.

## 2. Literature Review

This section focuses on the progress shown by relevant studies that have identified how intercountry integration affects electricity generation efficiency.

The existing literature has identified the positive effect of integration on efficiency in electricity generation. Regarding empirical works, the relationship has been measured in different contexts such as the European Union [9], Asia [10], South Africa [11], Central America [12], and South America, in particular, Chile, Argentina, and Peru [3]. However, the methodology used in these studies has been similar, focusing on the use of simulations that capture different scenarios.

Furthermore, integration can be associated not only with the augmentation of competition [3], but also with bolstering supply security [4,5] and enhancing efficiency in generation costs [3,6]. For example, Agostini et al. [3] demonstrates the viability of transparent and clear energy exchange between Chile and its neighbouring nations, thereby diminishing marginal energy costs and overall operational expenses, while maintaining a relatively consistent average generation cost. Moreover, Newbery and Grubb [4] showed that interconnectors facilitate the physical importation of power, providing supplementary alternatives to fulfil domestic demand. Consequently, it can be inferred that interconnectors generally contribute to the enhancement of supply security, except in exceedingly rare scenarios characterized by extreme combinations of contractual and physical factors. Finally,

Brinkerink et al. [6], based on a literature review, identified pivotal trends associated with the global grid concept, revealing a notable decline in costs pertaining to long-distance transmission technologies.

Furthermore, DG ENER [9] estimated the value of integration among European Union countries under different scenarios from 2015 to 2030 ranged from EUR 12.5 to EUR 40 billion, compared to the dependence on national autarky. For South Asia, Timilsina and Toman [10], there are estimated potential savings of up to USD 226 billion in the provision of electricity due to crossborder electricity trade between 2015 and 2040. In South America, Agostini et al. [3] simulated a regulatory framework for electricity trade between Chile, Argentina, and Peru, showing annual benefits of USD 40.6 million for the three countries because of the decrease in generation costs. In Central America, Echevarria et al. [12] estimated the benefit of the Electric Interconnection System for Central American Countries (SIEPAC) at USD 1409 billion between 2019 and 2025 in a higher integration scenario in the region. Finally, in South Africa, ESMAP [11] estimated the savings from full integration with a single market and only one system operator, ranging from $100 million to $1.5 trillion per year between 1997 and 2002. In general, integration between countries leverages the complementarities between them.

Other authors have evaluated the factors that affect productivity in the electricity generation sector. For example, Ajayi [13] found that greater vertical integration associated with natural monopolistic market structures, a higher degree of public ownership, and higher barriers to entry, are statistically significant in increasing generation costs in OECD countries.

Plant ownership is a widely studied variable in the literature. Studies in different countries have found that privately owned plants tend to be more efficient than publicly owned ones because of their focus on making profits [14–18]. However, others have found that the difference in plant efficiency is greater for public plants or is not significant according to the type of property [19,20]; in the short run, Pollitt, [18].

The literature also investigates the diversification of matrix generation as a variable of interest. Specifically, the examination focuses on analysing the complementarity among resources, which subsequently influences strategic decision making concerning technological choices in electricity generation [21,22]. Studies show that investment in different sources of electricity generation can be beneficial by exploiting each country's resources and taking advantage of geographical complementarities, reducing production costs, investment risks, and price volatility [21,22].

On the other hand, studies have focused on evaluating the impact of reforms in the electricity sector that seek to increase competition and encourage private investment in the sector. These studies identify the impacts of the reforms on installed capacity, labour productivity, and efficiency in electricity generation [23,24]. For example, Zhang et al. [23] found that competition among generators has positive effects on electricity generation, and electricity generation per worker is positively related to private participation within well-defined regulatory frameworks. In addition, Erdogdu [24] reveal that progress in the direction of electricity market reform is positively correlated with increased efficiency. Nevertheless, it is imperative to acknowledge that this effect is constrained and exhibits noteworthy variability across diverse country groups, evidencing the necessity to control at country-level. In this regard, the literature has substantiated that several variables, designed to control for country-specific effects, encompass factors such as industrialization, political rights, civil liberties, entry barriers, urbanization, climatic variables, electricity regulatory frameworks, and income [13,23–25]. In terms of tax of tariffs, according to Bustos [26], lower tariffs may induce foreign technology adoption.

## 3. Methodology

In this study, we analyse the determinants of efficiency in electricity generation for 24 countries in Latin America and the Caribbean for the 2000–2020 period. This section includes three subsections. The first contains the description of the data and the information.

The second presents the measurements used to construct the variables. Finally, the third presents the model used to evaluate the relationship between integrating electricity markets and efficiency in the electricity generation sector proposed in this study.

### 3.1. Data and Information Sources

The data collected comes from different information sources. We obtained the electricity generation efficiency level from the previous work of Cuadros et al. [8]. Imports, exports, and electricity generation from the different generation sources come from the U.S. Energy Information Administration [27]. We obtained the average tariff rate level for all imported products from the World Bank. The number of firms with public and private capital to build private participation comes from Sielac-Olade [28].

Finally, we obtained the year in which the country liberalized the electricity sector based on different sources of information. In Argentina, we consulted Resolution 240 of 2003 [29]; in Bolivia, Law 1604 of 1994 [30]; in Brazil, Law 9648 of 1998 [31]; in Chile, we consulted the General Law of Electric Services [32]; in Colombia, we considered the information of the CREG's website [33]; in the Dominican Republic, the Regulation for the application of the Electricity Law No. 125-01 [34]; in Ecuador, the Electrical Sector Regime Law [35]; in El Salvador, we took the information from the General Superintendence of Electricity and Telecommunications' website [36]; in Guatemala, the National Electric Power Commission [37]; in Honduras, the General Law of the Electricity Industry [38]; in Mexico, we considered the information from the National Energy Control Center [39]; in Nicaragua, we consulted the Electricity Industry Law [40]; in Panama, we obtained the in-formation from Law No. 6 of the Regulatory and institutional framework for the provision of public electricity service [41]; in Peru, Decree Law 25844 [42]; and, in Uruguay, Law 16832 [43].

In sum, Table 1 presents the description and sources of the variables.

**Table 1.** Variables and sources.

| Variable | Description | Source |
|---|---|---|
| Dependent variable | | |
| Efficiency of electricity generation | Own calculation based on the Dynamic DEA model | Cuadros et al. [8] |
| Variable of interest | | |
| Openness | Own calculation: Share of electricity exports and imports over the total electricity generation | The U.S. Energy Information Administration |
| Control variables | | |
| Diversification of the energy matrix | Own calculation: Diversification of the generation matrix based on the entropy concept | The U.S. Energy Information Administration |
| Ownership of plants | Own calculation: Participation of privately owned power plants in the generation sector | Sielac-Olade |
| Market liberalization | Own calculation using information on a country-level basis | Different sources of information on a country-level basis |
| Tariff rate | Unweighted average of the rates applied for all products subject to tariffs calculated for all traded goods | World Bank |

### 3.2. Measurements

In this section we describe the dependent variable (efficiency in electricity generation), the variable of interest, and, finally, the control variables.

3.2.1. Dependent Variable

The dependent variable is the efficiency of electricity generation that we obtained from the work of Cuadros et al. [8], based on a dynamic Data Envelopment Analysis (DEA) model. The DEA was proposed in Charnes et al. [44] and is one of the most important tools for measuring relative efficiency in different fields [45].

The model involves the generation of electricity as a desirable output; as undesirable output, the $CO_2$ emissions produced in the generation activity; and, as inputs, the *GDP* per capita and the installed capacity in the different generation sources. This score ranges from 0 to 1, indicating the minimum and maximum efficiency, respectively.

The model is based on the dynamic Data Envelopment Analysis (DEA) model of Constant Returns to Scale (CRS) based on slacks proposed by Tone and Tsutsui [46], extended to include the undesirable interdependence between electricity generation of electricity from fossil sources and $CO_2$ emissions, presented in Zhou and Liu [47].

The DEA model defines a production possibility frontier for each year based on the outputs and inputs of each country. We used the dynamic DEA model proposed by Cuadros et al. [8], whose formulation is based on Zhou and Liu [47] and that asserts the maximization of generation of electricity, and the minimization of $CO_2$ emissions can be achieved from an additive model by the Equation (1):

$$\max SDO\_NF_{ot} + SDO\_F_{ot} + SUO\_CO_{2_{ot}} \tag{1}$$

where $SDO\_NF_{ot}$ and $SDO\_F_{ot}$ refer to the inefficiencies of generation of electricity from nonfossil and fossil sources, respectively, and $SUO\_CO_{2ot}$ is the excesses of the emissions caused by the electricity generation activity. In the DEA literature, these inefficiencies and excesses are known as slacks.

The production possibility set for each country (country $o = 1, \ldots, 24$) under CRS is defined by Equations (2)–(10).

Equations (2)–(4) represent the constraints on inputs:

$$GDP_{ot} = \sum_{j=1}^{24} GDP_{jt}\lambda_j^t + S\_GDP_{ot} \tag{2}$$

where $\lambda_j^t$ is the linear combination coefficient of each country $j$ in year $t$. The *GDP* of country *"o"* must be less than or equal to the linear combination of the *GDP* of all countries in each year $t$. The difference is the slack variable of the *GDP* of country $o$ in year $t$ ($S\_GDP$).

$$IC\_NF_{ot} = \sum_{j=1}^{24} IC\_NF_{jt}\lambda_j^t + SIC\_NF_{ot} \tag{3}$$

The nonfossil installed capacity ($IC\_NF$) of country $o$ must be less than or equal to the linear combination of the nonfossil installed capacity of all countries in each year $t$. The difference is the slack variable of the $IC\_NF$ of country $o$ in year $t$ ($SIC\_NF$).

$$IC\_F_{ot} = \sum_{j=1}^{24} IC\_F_{jt}\lambda_j^t + SIC\_F_{ot} \tag{4}$$

The fossil installed capacity ($IC\_F$) of country $o$ must be less than or equal to the linear combination of the fossil installed capacity of all countries in each year $t$. The difference is the slack variable of the $IC\_F$ of country $o$ in year $t$ ($SIC\_F$).

Equations (5)–(6) represent the constraints of the electricity generation from nonfossil sources and fossil sources, respectively:

$$DO\_NF_{ot} = \sum_{j=1}^{24} DO\_NF_{jt}\lambda_j^t + SDO\_NF_{ot} \tag{5}$$

The generation of electricity from nonfossil sources ($DO\_NF$) of country $o$ must be less than or equal to the linear combination of electricity generation from the nonfossil sources of all countries in each year $t$. The difference is the slack variable of the $DO\_NF$ of country $o$ in year $t$ ($SDO\_NF$).

$$\alpha DO\_F_{ot} = \sum_{j=1}^{24} DO\_F_{jt}\lambda_j^t + SDO\_F_{ot} \tag{6}$$

The electricity generation from fossil sources ($DO\_F$) of country $o$ must be less than or equal to the linear combination of electricity generation from the fossil sources of all countries in each year $t$. The difference is the slack variable of the $DO\_F$ of country $o$ in year $t$ ($SDO\_F$). In Equation (6), $\alpha$ represents the interdependence between generation from fossil sources and the $CO_2$ emissions.

$$\alpha UO\_CO_{2ot} = \sum_{j=1}^{24} UO\_CO_{2jt}\lambda_j^t + SUO\_CO_{2ot} \tag{7}$$

The $CO_2$ emissions from the fossil sources ($UO\_CO_2$) of country $o$ must be less than or equal to the linear combination of $CO_2$ emissions of all countries in each year $t$. A variation in electricity generation from fossil sources is accompanied by the same proportional variation in $CO_2$ emissions.

Following the model proposed by Cuadros et al. [8], the level of installed capacity for each country in each year $t$ determines the installed capacity in the immediately succeeding year. Equations (8)–(9) represent this assumption:

$$\sum_{j=1}^{24} IC\_NF_{jt}\lambda_j^t = \sum_{j=1}^{24} IC\_NF_{jt}\lambda_j^{t+1} \tag{8}$$

$$\sum_{j=1}^{24} IC\_F_{jt}\lambda_j^t = \sum_{j=1}^{24} IC\_F_{jt}\lambda_j^{t+1} \tag{9}$$

Equation (10) captures the CRS assumption:

$$\sum_{j=1}^{24} \lambda_j^{t+1} \geq 0 \tag{10}$$

Additionally, non-negativity conditions are required:

$$S\_GDP_t,\ SDO\_NF_t,\ SDO\_F_t,\ SIC\_NF_t,\ SIC\_F_t,\ SUO\_CO2_t \geq 0 \tag{11}$$

Finally, for each country, we calculated the measure for each year ($\tau_{ot}^*$) and the overall efficiency measure ($\tau_o^*$), from the Equations (12)–(13), respectively:

$$\tau_{ot}^* = \frac{1}{1 + \frac{1}{3}\left(\frac{SDO\_NF_{it}}{DO\_NF_{it}} + \frac{SDO\_F_{it}}{DO\_F_{it}} + \frac{SUO\_NF_{it}}{UO\_CO_{2it}}\right)};\ t = 2000, \ldots, 2020 \tag{12}$$

$$\tau_o^* = \frac{1}{21}\sum_{2000}^{2020} \tau_{otit}^*;\ t = 2000, \ldots, 2020 \tag{13}$$

a country $o$ will be overall efficient $\tau_o^* = 1$ if and only if is efficient in each year.

From Equation (12), a country with input or $CO_2$ emission excess will be an inefficient country, and will have an overall efficiency measure less than 1, given by the simple average of the efficiency measure year over year as shown in Equation (13).

In sum, the dynamic DEA model measures efficiency in generation of electricity for each country and year, accounting for inefficiencies in generation or excesses in inputs.

In the work of Cuadros et al. [8], they cover the 2000–2016 period. We used the same formulation of Dynamic DEA, we updated the data and we covered the period from 2000 to 2020.

### 3.2.2. Variable of Interest

The variable of interest is the level of openness in the electricity sector, which captures the integration level, and it is calculated based on the imports and exports level as a proportion of electricity generation. This variable is proposed considering that interconnections provide benefits as security of supply [4,5], reducing the generation costs [3,6], increasing competition [3], and decreasing environmental costs [3].

The following is the definition of the country-level trade openness metric:

$$\frac{x+m}{GDP} \quad (14)$$

where $x$ refers to exports, $m$ refers to electricity imports, and $GDP$ is the gross domestic product.

However, to capture the openness of the generation sector, we propose to calculate the share of electricity exports and imports over the total electricity generation in country $i$ in year $t$, following the mathematical expression:

$$Openness_{it} = \left(\frac{X_{eit} + M_{eit}}{EG_{it}}\right) * 100 \quad (15)$$

where $X_{eit}$ refers to electricity exports, $M_{eit}$ represents the electricity imports, and $EG_{it}$ is the electricity generation in country $i$, year $t$.

Thus, a country in complete autarky in the electricity sector presents an openness level of zero, for example, island countries. Countries with relatively high volumes of electricity trade are those with the highest levels of openness.

### 3.2.3. Control Variables

For both industries and nations, we propose four control variables. We suggest diversifying the energy matrix and establishing a private ownership percentage for the companies engaged in generation activity at the industry level. Nation-wise, we use the liberalization of the electricity market and the tax rate. We explain each of the variables below.

- Diversification of the Energy Matrix

The literature has evaluated the benefit of diversification in terms of risk reduction and not as a competitiveness strategy. In this sense, Garcia-Mazo [21] analysed the diversification of the electricity generation matrix from financial and strategic points of view. Greater diversification represents a better response to the uncertainty caused by fluctuating fuel prices and climate, as well as a more efficient use of natural resources and a lower environmental. Consequently, investing in different generation sources with seasonal and geographic complementarities not only reduces investment risks, costs, and price volatility, but also increases the benefits of generation activity [22].

We carried the measurement of diversification in the generation sector, using the entropy index [48]. The Shannon entropy index used is derived from a previous measure as a measure of dispersion in a distribution [49].

We used the Shannon entropy index for country $i$ in year $t$ as a proxy for the diversification of the generation matrix and we calculated it from the following expression:

$$Diversification_{it} = e^{-\sum_{j=1}^{7} p_{ijt} \ln p_{ijt}} \quad (16)$$

where $p_{ijt}$ represents the share of generation source of each of the seven sources $j$ over the total generation of country $i$ in year $t$. In our case, we the information disaggregated in seven generation sources: Nuclear, fossil fuels, hydroelectricity, geothermal, solar (or tide, wave, and fuel cell), wind, and biomass (and waste). Therefore, the value range for diversification is 1 to 7, where 1 denotes a country whose generation matrix is entirely concentrated in a single generation source and 7 denotes a country whose generation sources have the same participation in its generation matrix (1/7).

- Plant Ownership

Theoretically, competition in the electricity sector generates gains in terms of efficiency and productivity. This relationship has been evaluated in many empirical studies and most have found, controlling different factors, that privately owned plants are more efficient than publicly owned plants because they have better incentives to maximize profits rather than focusing on public services [14–17] and, in the long run, Pollit [18].

We calculated the participation of privately owned power plants in the generation sector in country $i$ from the following expression:

$$Privately\ owned_i = \frac{\#\ private\ plants_i}{\#\ total\ plants_i} \tag{17}$$

where the numerator is the number of private plants in country $i$, at the end of the period, and the denominator is the total number of plants in country $i$, at the end of the period.

- Market Liberalization

State companies dominated the Latin American electricity systems before the reforms. These companies were vertically integrated for the generation, transmission, and distribution of electricity. Until the early 1990s, governments owned the electricity sector or controlled it through strict regulations. World reforms in the late 1980s aimed to break the idea of electricity as a natural monopoly, promoting competition and private investment in the sector [50].

We considered the year of market liberalization in the countries involved in the study. We used a dummy variable to indicate the year the electricity market in country $i$ in year $t$ was liberalized:

$$M_{it} = \begin{cases} 1 & if\ the\ electricity\ market\ is\ liberalized\ in\ the\ year\ t \\ 0 & otherwise \end{cases} \tag{18}$$

- Tariff Rate

The simple average applied tariff was calculated as the unweighted average of the rates applied for all products subject to tariffs calculated for all traded goods. In this definition, tariff is a tax on goods and services imported and exported into a country. We used this measure as a proxy of willingness to adopt new technologies. Firms in countries with lower tariffs have incentives to import and adopt new technology from other countries.

The World Bank estimates this measure using the world integrated trade solution system from the World Trade Organization database.

### 3.3. Model Specifications

A hurdle model was used to identify the determinants of efficiency in electricity generation. This statistical model was used when the dependent variable was bounded on the left and on the right and showed an excess of concentration of the data over a lower or upper limit.

The hurdle model assumes that the censored observations are independent of the uncensored ones, and it was proposed by Cragg [51]. The model estimates two equations: one for the censored situation and another to explain the uncensored data. The model for censored data involved the analysis of the probability of attaining value 1 or 0 (binary model), and the model for uncensored data models, the probability of nonzero values (counting model). The model estimation was performed using maximum likelihood.

The mathematical formulation of the hurdle model in our study is as follows:

### 3.4. Binary Model: (Model of Efficient Countries)

$$s_i = \begin{cases} 1\ if\ z_i\gamma + u_i \geq 1 \\ 0\ if\ z_i\gamma + u_i < 1 \end{cases} \tag{19}$$

where $s_i$ is a dummy variable that takes the value of 1 if the efficiency of electricity generation for country $i$ is 1, and takes the value of 0 otherwise, $z_i$ is a vector of independent variables, $\gamma$ is a vector of coefficients of independent variables, and $u_i$ is the error term distributed normally ($u_i \sim N(0,1)$).It is important to note that $s_i = 1$ is estimated using a Probit model, whose mathematical formulation is as follows:

$$P(s_i = 1) = z_i\gamma + u_i \quad (20)$$

Therefore, we estimate the variables that determine whether a country is efficient, using the Probit model proposed in (8).

*3.5. Counting Model: (Model of Nonefficient Countries)*

$$Y_i^* = x_i\beta + \varepsilon_i \quad (21)$$

$$Y_i = Y_i^* \; if \; s_i = 0 \; and \; Y_i^* > 0 \quad (22)$$

where $x_i$ is a vector of explanatory variables, $\beta$ is a vector of coefficients, and $\varepsilon_i \sim N(0,\sigma^2)$. It is estimated by using a truncated normal regression. $s_i$ is a dummy variable that takes the value of 0 if the efficiency of electricity generation for country $i$ is less than 1.

Therefore, we estimated the variables that determine whether a country presents efficiency of electricity generation less than 1, using the model formulated in (9).

In summary, we employed a two-step model known as a hurdle model. In the first model, we analysed the determinants of whether a country is efficient in electricity generation. In the second, we analysed the determinants of countries that are inefficient in electricity generation activity. Therefore, we model the efficiency of electricity generation in two steps.

Our data followed a panel structure. Due to the computational difficulty of including the panel data structure in hurdle models, we tested the temporal homoscedasticity hypothesis between countries that may affect the efficiency of regression coefficients estimated. The used test is a Z test and was contrasted against a normal distribution. The null hypothesis states that no variable affects the conditional variance in each model. In our case, we tested if the country variable affects the conditional variance in both the bi-nary and the counting models.

## 4. Results and Their Discussion

This section contains two parts. The first shows the descriptive statistics of the variables used in the proposed model. The second presents the results obtained from the estimated model.

*4.1. Descriptive Analysis*

Table 2 presents the descriptive statistics of the data set used. We classified the countries in three groups in relation to their overall efficiency levels.

**Table 2.** Descriptive statistics.

| Country | Statistic | | Dependent Variable | Independent Variables | | | | |
|---|---|---|---|---|---|---|---|---|
| | | Efficiency Group | Efficiency | Openness | Diversification | Liberalization | Privately Owned | Tariff |
| AR | Mean | 1 | 0.90 | 9.96 | 2.42 | 0.86 | 0.55 | 13.01 |
| | SD | | 0.03 | 2.83 | 0.20 | 0.36 | | 0.93 |
| BO | Mean | 2 | 0.83 | 0.06 | 2.10 | 1.00 | 0.58 | 10.39 |
| | SD | | 0.07 | 0.12 | 0.13 | 0.00 | | 1.39 |
| BR | Mean | 1 | 1.00 | 8.26 | 2.27 | 1.00 | 0.73 | 13.51 |
| | SD | | 0.00 | 2.39 | 0.53 | 0.00 | | 0.68 |

**Table 2.** *Cont.*

| Country | Statistic | Efficiency Group | Dependent Variable | Independent Variables | | | | |
|---|---|---|---|---|---|---|---|---|
| | | | Efficiency | Openness | Diversification | Liberalization | Privately Owned | Tariff |
| CH | Mean<br>SD | 1 | 0.93<br>0.04 | 1.75<br>1.74 | 2.62<br>0.45 | 1.00<br>0.00 | 0.7 | 6.28<br>0.81 |
| CO | Mean<br>SD | 1 | 0.91<br>0.05 | 1.74<br>1.11 | 1.87<br>0.15 | 1.00<br>0.00 | 0.16 | 9.65<br>3.03 |
| CR | Mean<br>SD | 3 | 0.33<br>0.24 | 6.85<br>3.25 | 2.27<br>0.30 | 0.00<br>0.00 | 0.18 | 5.74<br>0.33 |
| CU | Mean<br>SD | 1 | 1.00<br>0.00 | 0.00<br>0.00 | 1.24<br>0.05 | 0.00<br>0.00 | 0 | 10.58<br>0.29 |
| DR | Mean<br>SD | 2 | 0.75<br>0.11 | 0.00<br>0.00 | 1.54<br>0.14 | 0.90<br>0.30 | 0.32 | 8.34<br>2.51 |
| EC | Mean<br>SD | 1 | 0.88<br>0.07 | 4.81<br>4.28 | 2.05<br>0.14 | 1.00<br>0.00 | 0.34 | 12.25<br>1.73 |
| ES | Mean<br>SD | 2 | 0.74<br>0.10 | 13.23<br>10.33 | 3.52<br>0.55 | 1.00<br>0.00 | 0.06 | 6.32<br>0.58 |
| GU | Mean<br>SD | 2 | 0.80<br>0.06 | 10.55<br>7.48 | 3.16<br>0.30 | 1.00<br>0.00 | 0.69 | 5.83<br>0.47 |
| GY | Mean<br>SD | 3 | 0.68<br>0.08 | 0.00<br>0.00 | 1.29<br>0.05 | 0.00<br>0.00 | 0.23 | 11.59<br>2.30 |
| HA | Mean<br>SD | 3 | 0.54<br>0.11 | 0.00<br>0.00 | 1.74<br>0.25 | 0.00<br>0.00 | 0 | 3.91<br>1.10 |
| HO | Mean<br>SD | 2 | 0.83<br>0.07 | 4.29<br>3.51 | 2.80<br>0.98 | 0.33<br>0.48 | 0.74 | 5.86<br>0.48 |
| JA | Mean<br>SD | 2 | 0.75<br>0.15 | 0.00<br>0.00 | 1.36<br>0.19 | 0.00<br>0.00 | 0.23 | 8.01<br>1.12 |
| MX | Mean<br>SD | 1 | 1.00<br>0.00 | 1.15<br>0.99 | 2.09<br>0.13 | 0.24<br>0.44 | 0 | 11.65<br>4.51 |
| NI | Mean<br>SD | 2 | 0.72<br>0.06 | 3.58<br>5.34 | 3.23<br>0.96 | 0.81<br>0.40 | 0.48 | 5.42<br>0.51 |
| PN | Mean<br>SD | 2 | 0.81<br>0.09 | 2.78<br>1.40 | 2.12<br>0.23 | 0.57<br>0.51 | 0.2 | 7.20<br>0.70 |
| PY | Mean<br>SD | 1 | 1.00<br>0.00 | 80.07<br>7.89 | 1.01<br>0.01 | 0.00<br>0.00 | 0 | 10.74<br>1.07 |
| PE | Mean<br>SD | 1 | 0.96<br>0.09 | 0.05<br>0.08 | 2.06<br>0.28 | 1.00<br>0.00 | 0.49 | 6.70<br>3.97 |
| SU | Mean<br>SD | 2 | 0.73<br>0.08 | 11.74<br>17.38 | 2.00<br>0.08 | 0.00<br>0.00 | 0 | 11.05<br>2.28 |
| TT | Mean<br>SD | 3 | 0.51<br>0.27 | 0.00<br>0.00 | 1.01<br>0.01 | 0.00<br>0.00 | 0 | 8.98<br>1.30 |
| UR | Mean<br>SD | 3 | 0.49<br>0.30 | 17.82<br>12.93 | 2.27<br>0.83 | 1.00<br>0.00 | 0.04 | 11.09<br>1.19 |
| VE | Mean<br>SD | 1 | 0.99<br>0.02 | 0.68<br>0.36 | 1.87<br>0.08 | 0.00<br>0.00 | 0.09 | 13.14<br>0.53 |

Source: Own elaboration. Labels: AR: Argentina, BO: Bolivia, BR: Brazil, CH: Chile, CO: Colombia, CR: Costa Rica, CU: Cuba, DR: Dominican Rep., EC: Ecuador, ES: El Salvador, GU: Guatemala, GY: Guyana, HA: Haiti, HO: Honduras, JA: Jamaica, MX: Mexico, NI: Nicaragua, PN: Panama, PY: Paraguay, PE: Peru, SU: Suriname, TT: Trinidad and Tobago, UR: Uruguay, VE: Venezuela.

In Table 2, the third column represents the efficiency group. We considered interesting to analyse the descriptive statistics for each of them using different maps for each variable.

Figure 1 shows the spatial distribution of the analysed variables using quantile maps.

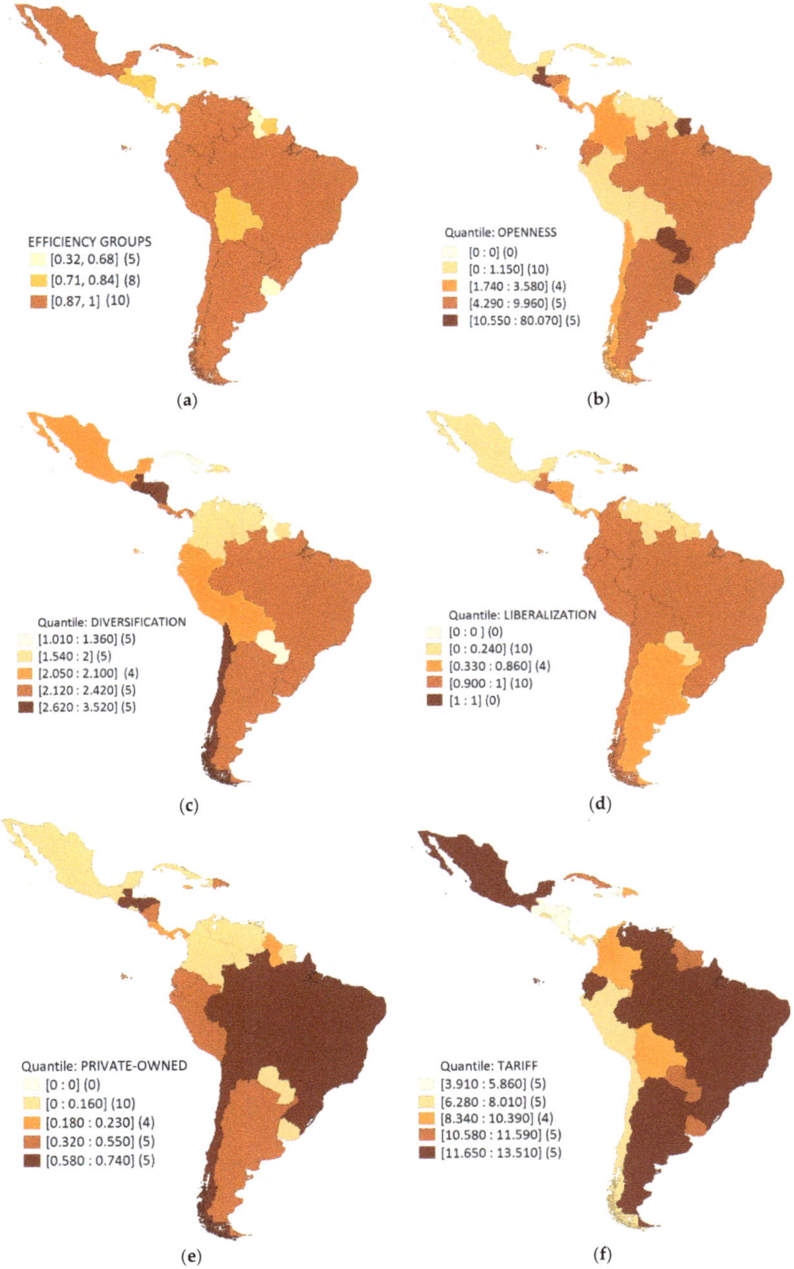


**Figure 1.** Quantile maps of analysed variables. Figure 1 contains the quantile maps of the variables used in the analysis, respectively: (**a**) Efficiency groups, (**b**) openness levels, (**c**) diversification values, (**d**) liberalization, (**e**) private owned participation (**f**) tariffs rates.

Regarding efficiency, group 1 is constituted by 10 countries with overall efficiencies that vary between 0.87 and 1. These countries are: Brazil, Cuba, Mexico, Paraguay, Venezuela, Peru, Chile, Colombia, Argentina, and Ecuador. Cuba presents value of zero in terms of openness, which can be interpreted as that the electricity demand of this country is completely covered by its generation. This country presents the second lowest level of diversification of the countries of the group, and their power system does not have free competition, where all generation plants are publicly owned. In addition, countries like Venezuela, Mexico, Colombia, Chile, and Peru present values of openness higher than 0 and lesser than 1.75%, evidencing low energy exchanges with other countries. These countries stand out for being countries that diversify their generation with at least two sources. Of these countries, Venezuela is the only one that does not present free competition in the electricity market. On the other hand, Ecuador, Brazil, and Argentina present openness values between 4.8% and 10%, whose diversification is higher than 2, meaning that their generation is covered by at least three sources, and whose electricity markets are characterized by having been liberalized before the first year considered in the analysis period and since 2003 in the case of Argentina. Paraguay is the country with the highest level of openness, its exchange of energy with other countries is about 80%, but with a diversification level of 1, showing that the country's generation matrix is almost absolutely oriented towards one only source: hydroelectricity. In addition, its electricity market has not been liberalized and all generation plants in the country are state-owned. Finally, the level of taxes applied to imported products in the countries of group 1 was 10.75% on average between 2000 and 2020. The highest tariff level was Brazil with 13.51%, while the minimum observed was Chile with 6.28%.

The second efficiency group is made up of nine countries and their overall efficiency levels vary between 0.71 and 0.84. These countries are: Bolivia, Honduras, Panama, Guatemala, Jamaica, Dominican Republic, El Salvador, Suriname, and Nicaragua. Of these countries, Jamaica, Dominican Republic, and Bolivia have null or almost null exchanges of electricity, since their openness values in the period analysed are 0 in the two first cases, and 0.06 in the Bolivian case. The range of the diversification measure of these countries is between 1.3 and 2.1 and implies that their generation involves two sources, in Jamaica and Dominican Republic and three sources in Bolivia.

In addition, Suriname and Jamaica are the only countries that do not present free competition in the electricity market. Regarding Suriname, it has the second largest exchange of electricity in this group, with an 11.74% openness level.

Furthermore, Panama and Nicaragua have two aspects in common: (i) Their power systems were not subject to free competition from the beginning of the analysed period, and (ii) their plants are mostly publicly owned. In addition, the generation of electrical energy in Nicaragua depends on three sources, while in Panama it depends on two generation sources.

Finally, the bordering countries—Honduras, Guatemala, and El Salvador—showed openness levels between 4.29% and 13.33%. Moreover, their diversification indexes were between 2.8 and 3.6, implying that their generation matrices involve between three and four sources. In addition, Guatemala and El Salvador have had a free competitive power system before the analysed period, unlike Honduras, which only allowed free competition since 2014. Regarding plant ownership, almost all of them are publicly owned in El Salvador, while most are privately owned in Honduras and Guatemala. Lastly, the average level of taxes to imported goods in the nine countries was similar to the mean observed in 24 countries in the study, 8.91% compared with 9.19% for the 24 countries. The lowest tariff level was Nicaragua, with 5.42%, while the highest observed level was Suriname with an 11.05%.

The overall efficiency of the group of countries with the lowest generation efficiency levels varies between 0.32 and 0.68, consisting of Guyana, Haiti, Trinidad and Tobago, Uruguay, and Costa Rica. Of these countries, Haiti, Trinidad and Tobago, and Guyana have three things in common: (i) They do not have exchanges of electricity with other countries, showing values of zero in terms of openness, (ii) they use two sources of generation in their matrix, (iii) the electricity systems in these countries do not have free competition. On the other hand, all the plants of Haiti and Trinidad and Tobago are publicly owned, and, in Guyana, this participation is around the 77%. Regarding Uruguay and Costa Rica, they have openness levels of 17.82 and 6.85, respectively, being Uruguay the second largest extent of this index out of the 24 countries. Both countries generate electricity from three sources and their plants are mainly publicly owned. However, they differ because Uruguay has had a free competition electricity system before 2000 and Costa Rica does not have free competition in the analysed period. Finally, the level of taxes of imported goods is larger than 11.1% in Trinidad and Tobago, Uruguay, and smaller than 6% in Costa Rica, Guyana, and Haiti.

Below, we present the correlation matrix for the independent variables, which helps examine their relationship and detect potential multicollinearity problems in the estimations.

Table 3 presents the correlation matrix of the independent variables. Due to all the correlations are less than 0.85, the independent variables do not present problems associated with multicollinearity [52,53].

**Table 3.** Correlation matrix of independent variables.

| Variable | Openness | Diversification | Private Ownership |
|---|---|---|---|
| Openness | | | |
| Diversification | −0.1204 | | |
| Private Ownership | −0.1948 | 0.4318 | |
| Tariff | 0.1310 | −0.3218 | −0.1246 |

### 4.2. Model Estimation

We used the hurdle model proposed by Cragg to identify the determinants of generation efficiency in 24 countries of Latin America and the Caribbean during the 2000–2020 period. It is important to note our data varied not only by country (i) but also by year (t).

We used linear and nonlinear factors to define the factors that determine the level of generation efficiency. We incorporated the linear factors into the analysis from the examined literature. The nonlinear factor was diversification because policymakers may be interested in identifying the inflection points that allow defining policies requiring transitions to these diversification levels.

In our study, the binary and counting models were given by:

Binary Model: (Efficient Countries Model)

Following the Probit model defined in (8), in our case, the binary model was given by:

$$P(s_{it}=1)=\gamma_0+\gamma_1 Openness_{it}+\gamma_2 Diversification+\gamma_3 Diversification_{it}^2+\gamma_4 M_{it}+\gamma_5 PrivateOwn_{it}+ \gamma_6 Tariff_{it}+u_{it} \quad (23)$$

where $s_{it}$ is a dummy variable that takes the value 1 whether country $i$ in year $t$ is efficient in its generation activity and takes the value of 0 otherwise, $\gamma_i$ is the vector of coefficients of independent variables, and $u_{it}$ is the error for country $i$ in year $t$ term distributed as a standard normal with mean zero and standard deviation 1.

Counting Model: (Nonefficient Countries Model)

Following (9) and (10), the counting model is defined as:

$$Y_{it}=\beta_0+\beta_1 Openness_{it}+\beta_2 Diversification+\beta_3 Diversification_{it}^2+\beta_4 M_{it}+\beta_5 PrivateOwn_{it}+ \beta_6 Tariff_{it}+\varepsilon_{it} \quad (24)$$

where $Y_{it}$ represents the efficiency of electricity generation for country $i$ in year $t$ for countries whose efficiency is higher than 0 and less than 1, $\beta_i$ is a vector of coefficients of the independent variables, and $\varepsilon_{it}$ is the error term for country $i$ and year $t$, which is distributed as $\varepsilon_{it}\sim N(0,\sigma^2)$. It is estimated by using a truncated normal regression.

Table 4 presents the hurdle model estimates. The left-hand side shows the estimates for the selection model, while the right-hand side contains the estimates for the uncensored or counting model.

**Table 4.** Binary and counting models.

| Variable | Binary Model | | Counting Model |
|---|---|---|---|
| | $\gamma$-Coefficients | | $\beta$ |
| Constant | 1.057 | | −0.277 |
| Openness | 0.087 ** | | −0.051 * |
| Diversification | −8.750 | | 1.909 |
| Diversification$^2$ | 1.431 | | −0.420 |
| Market liberalization | −0.871 | | 2.523 *** |
| Private Ownership | 3.102 | | 4.790 *** |
| Tariff | 0.751 ** | | 0.184 |
| $\Sigma$ | | | |
| Country | | 0.121 *** | |
| $\Omega$ | | | |
| Country | | 0.008 | |

$p$-values: *** 0.01, ** 0.05, * 0.1.

Table 4 shows the estimation for the binary and counting models. We estimated the model fitting a heteroskedastic Probit using the identifier of the country for modelling the conditional variance in both the binary and counting models. We tested the null hypothesis of homoscedasticity between countries, and we found that this variable is only statistically significative for the Probit model.

We tested the null hypothesis of homoscedasticity between countries. $\sum$ represents the information about the estimated standard deviation of the error term in the selection model and $\Omega$ the information about the counting model. Results showed that the country variable is significant at 1% and there is evidence that the countries affect the variance of the selection model. On the other hand, there is no evidence that countries affect the variance of the counting model.

Regarding the results of the binary model, there is a higher tendency to be completely efficient when facing increases in openness in the electricity sector and increases in the generalized level of taxes in the economy, with significant effects occurring at the 5% level. Additionally, the type of property, the matrix diversification, and electricity market liberalization do not have a significant impact on the probability of observing an efficient result for a single year.

On the other hand, because the hurdle model is a nonlinear model, the coefficients of the counting model cannot be interpretable directly. Table 5 presents the marginal effects of each variable of the counting model.

**Table 5.** Marginal effects.

| Variable | $dy/dx$ |
|---|---|
| Openness | −0.001 |
| Diversification | −0.050 |
| Diversification$^2$ | 0.003 |
| Market liberalization | 0.104 ** |
| Private Ownership | 0.273 *** |
| Tariff | 0.020 *** |

$p$-values: *** 0.01, ** 0.05.

The results show that, for countries with uncensored efficiency levels lower than 1, openness in the electricity sector does not have a significant impact on the level of efficiency, which is contrary to what was found in different regions around the world based on simulation of scenarios that capture a total integration between countries. For example, DG ENER [9] estimated the value of integration for the countries of the European Union between 2015 and 2030. Timilsina and Toman [10] did the same for the countries of South Asia between 2015 and 2040. In South America, Agostini et al. [3] simulated a regulatory framework for electricity trading between Chile, Argentina, and Peru. In Central America, Echevarria et al. [12] estimated the benefit of the SIEPAC between 2019 and 2025. Finally, in South Africa, ESMAP [11] estimated the savings from a complete integration, a single market and only one system operator between 1997 and 2002. All these studies pointed to important benefits of integration in terms of savings in electricity generation due to cross-border trade.

We did not find a significant impact of the diversification of the generation matrix on the efficiency of electricity generation. As far as we know, only Perez-Odeh [22] argued the existence of a positive relationship between diversification and efficiency.

The results suggest that the liberalization of the electricity market has positive and significant effects at 1% on efficiency. The liberalization of the market has an average impact of 0.104 on the efficiency level exhibited regarding efficiency in years when the market is not liberalized.

Private participation in the generation of electricity has positive and significant effects at 1%. Increases of 0.1 in the participation of private plants increase generation efficiency by 0.027 ($0.273 \times 0.1 = 0.027$). This result follows the same path found in many studies that use the generation plant as the unit of analysis and find that those with private capital are more efficient than those with public capital in different countries. For example, See and Coelli [14] studied the efficiency of thermal plants in Malaysia between 1998 and 2005. Sarica and Or [15] included thermal, hydroelectric, and wind plants in Turkey. Khanna et al. [16] worked with fossil and thermal plants in India. Dean [17] analysed the efficiency of coal and natural gas plants in the United States, while Pollit [18] used different plants around the world, finding that long-term private plants are more efficient than public ones.

Finally, there is a positive and significant impact of tariff rate on efficiency. In other words, if the level of taxes in the economy increases 1%, the efficiency in the inefficient countries increases 0.02. Protecting local industry seems to have a positive impact on electricity generation efficiency in the region, contradicting Bustos' result [26]. Olcay and Laing [54] found that governments can benefit and protect local industries from protectionist policies, like the pharmaceutical industry.

## 5. Conclusions

In this work, we evaluated the determinants of electricity generation efficiency in 24 Latin America and Caribbean countries during the 2000–2020 period using the proposed model by Cuadros et al. [8], derived from the DEA methodology. Our variable of interest measured the openness in the electricity sector as the share of exports and imports over the total generation in each country. The methodology chosen—based on the censored characteristic of the dependent variable—was the hurdle model proposed by Cragg [51]. This methodology allowed an independent analysis of fully efficient countries from those that showed some inefficiency in the analysed period.

From the methodology, we can separate the determinants of the efficiency of electricity generation for efficient countries and for the inefficient ones.

Regarding the efficient countries, the results suggest a positive and significant openness effect in the electricity sector in the sample, such as Brazil, Mexico, and Paraguay. The Cuba case is distinct since it lacks electricity interconnections.

For inefficient countries, we did not find evidence of a significant openness effect on efficiency. On the other hand, we found evidence that market liberalization had the strongest effect on efficiency gains. The liberalization of the market has an average impact of 0.104 on the efficiency level exhibited regarding efficiency in years when the market is not liberalized. Therefore, we consider that countries like Costa Rica, Guyana, Haiti,

Jamaica, Suriname, and Trinidad and Tobago, could benefit from a liberalized electricity market in terms of greater energy efficiency. Following to Zhang et al. [23], inefficient countries can re-evaluate their regulatory electricity frameworks, taking the efficient ones as references.

In addition, we observed that participation of private plants has positive effects on efficiency for the inefficient countries. Increases of 0.1 in the participation of private plants increased generation efficiency by 0.023. In countries with low private ownership, such as Costa Rica, El Salvador, Guyana, Haiti, Jamaica, Panama, Suriname, Trinidad and Tobago, and Uruguay. We consider that this outcome is useful for policy makers to propose strategies focused on promotion of private investment in electricity generation sector.

Finally, the protectionism level of tariffs has positive effects for completely efficient and inefficient countries. Baldwin [55] proposes a theoretical model where the tariffs of the countries are noncooperative strategies, which leads to a suboptimal Nash equilibrium. The author found that unilateral tariff reductions were detrimental to the outcome of each country.

The evidence we present is important for the countries that are at the lowest level of efficiency—Haiti, Trinidad and Tobago, Guyana, Uruguay, and Costa Rica. Of these countries, Uruguay is the only one that has a liberalized electricity market. Moreover, all of them have low participation of private companies in electricity generation. We recommend that policymakers in these countries encourage private investment in the generation sector and implement policies aimed at liberalizing the electricity market. The regulatory frameworks of the most efficient countries provide elements that help improve their efficiency of electricity generation.

Further research could concentrate on evaluating joint reduction tariff policies and their effects on efficiency generation. Additionally, future research might aim to determine whether electricity prices in the region are converging. It could also assess the role of market integration in Latin American countries in shaping this trend.

**Author Contributions:** Conceptualization, Y.E.R.; Methodology, S.C.; Software, S.C.; Validation, S.C.; Formal analysis, S.C. and Y.E.R.; Data curation, S.C.; Writing—original draft, S.C.; Writing—review and editing, Y.E.R. and J.C.; Supervision, Y.E.R. and J.C. All authors have read and agreed to the published version of the manuscript.

**Funding:** This research received no external funding.

**Data Availability Statement:** Data are contained within the article.

**Conflicts of Interest:** The authors declare no conflict of interest.

## References

1. Cepeda, M. Assessing cross-border integration of capacity mechanisms in coupled electricity markets. *Energy Policy* **2018**, *119*, 28–40. [CrossRef]
2. Singh, A.; Jamasb, T.; Nepal, R.; Toman, M. Electricity cooperation in South Asia: Barriers to cross-border trade. *Energy Policy* **2018**, *120*, 741–748. [CrossRef]
3. Agostini, C.A.; Guzmán, A.M.; Nasirov, S.; Silva, C. A surplus based framework for cross-border electricity trade in South America. *Energy Policy* **2019**, *128*, 673–684. [CrossRef]
4. Newbery, D.; Grubb, M. Cambridge Working Papers in Economics the Final Hurdle? *Security of Supply, the Capacity Mechanism and the Role of Interconnectors*. 2014. Available online: www.eprg.group.cam.ac.uk (accessed on 20 January 2023).
5. Cepeda, M.; Saguan, M.; Finon, D.; Pignon, V. Generation adequacy and transmission interconnection in regional electricity markets. *Energy Policy* **2009**, *37*, 5612–5622. [CrossRef]
6. Brinkerink, M.; Gallachóir, B.; Deane, P. A comprehensive review on the benefits and challenges of global power grids and intercontinental interconnectors. *Renew. Sustain. Energy Rev.* **2019**, *107*, 274–287. [CrossRef]
7. Mulder, M.; Giesbertz, P. Economics of Interconnection: The Case of the Northwest European Electricity Market. 2008. Available online: https://www.researchgate.net/publication/228901114 (accessed on 20 May 2023).
8. Cuadros, S.; Rodríguez, Y.E.; Contreras, J. Dynamic data envelopment analysis model involving undesirable outputs in the electricity power generation sector: The case of latin america and the caribbean countries. *Energies* **2020**, *13*, 6624. [CrossRef]
9. Directorate-General for Energy. *Benefits of an Integrated European Energy Market*; DICE: Düsseldorf, Germany, 2013.

10. Timilsina, G.R.; Toman, M. Potential gains from expanding regional electricity trade in South Asia. *Energy Econ.* **2016**, *60*, 6–14. [CrossRef]
11. Energy Sector Management Assistance Program. Regional Power Sector Integration: Lessons from Global Case Studies and a Literature Review (Vol. 11): South African Power Pool (SAPP): Transmission & Trading Case Study (English). 2009. Available online: http://documents.worldbank.org/curated/en/126211468323330432/South-African-Power-Pool-SAPP-transmission-trading-case-study (accessed on 12 October 2023).
12. Echevarría, C.; Jesurun-Clements, N.; Mercado, J.; Trujillo, C. *Central American Electricity Integration: Genesis, Benefits and Outlook of the SIEPAC Project*; Inter-American Development Bank: Washington, DC, USA, 2017.
13. Ajayi, V.A. Essays on Deregulation in the Electricity Generation Sector. 2017. Available online: https://repository.lboro.ac.uk/articles/thesis/Essays_on_deregulation_in_the_electricity_generation_sector/9492179 (accessed on 29 September 2023).
14. See, K.F.; Coelli, T. An analysis of factors that influence the technical efficiency of Malaysian thermal power plants. *Energy Econ.* **2012**, *34*, 677–685. [CrossRef]
15. Sarica, K.; Or, I. Efficiency assessment of Turkish power plants using data envelopment analysis. *Energy* **2007**, *32*, 1484–1499. [CrossRef]
16. Khanna, M.; Mundra, K.; Ullah, A. Parametric and semi-parametric estimation of the effect of firm attributes on efficiency: The electricity generating industry in India. *J. Int. Trade Econ. Dev.* **1999**, *8*, 419–436. [CrossRef]
17. Hiebert, L.D. The Determinants of the Cost Efficiency of Electric Generating Plants: A Stochastic Frontier. 2002. Available online: http://www.jstor.org/stable/1061501 (accessed on 10 September 2023).
18. Pollitt, M. *Ownership and Performance in Electric Utilities*; Oxford University Press: Oxford, UK, 1995.
19. Pollitt, M.G. Ownership and Efficiency in Nuclear Power Production. *Oxf. Econ. Pap.* **1996**, *48*, 342–360. [CrossRef]
20. Fare, R.; Grosskopf, S.; Pasurka', C. Effects on relative efficiency in electric power generation due to environmental controls. *Resour. Energy* **1986**, *8*, 167–184. [CrossRef]
21. Mazo, C.M.G.; Olaya, Y.; Botero, S.B. Investment in renewable energy considering game theory and wind-hydro diversification. *Energy Strategy Rev.* **2020**, *28*, 100447. [CrossRef]
22. Odeh, R.P.; Watts, D.; Negrete-Pincetic, M. Portfolio applications in electricity markets review: Private investor and manager perspective trends. *Renew. Sustain. Energy Rev.* **2018**, *81*, 192–204. [CrossRef]
23. YZhang, F.; Parker, D.; Kirkpatrick, C. Electricity sector reform in developing countries: An econometric assessment of the effects of privatization, competition and regulation. *J. Regul. Econ.* **2008**, *33*, 159–178. [CrossRef]
24. Erdogdu, E. What happened to efficiency in electricity industries after reforms? *Energy Policy* **2011**, *39*, 6551–6560. [CrossRef]
25. Yu, B.; Fang, D.; Meng, J. Analysis of the generation efficiency of disaggregated renewable energy and its spatial heterogeneity influencing factors: A case study of China. *Energy* **2021**, *234*, 121295. [CrossRef]
26. Bustos, P. American Economic Association Trade Liberalization, Exports, and Technology Upgrading: Evidence on the Impact of MERCOSUR on Argentinian Firms. 2011. Available online: http://www.jstor.org/stable/41038790 (accessed on 20 May 2023).
27. EIA. Energy Information Administration. Available online: https://www.eia.gov/international/data/world (accessed on 12 October 2023).
28. Sielac-Olade. "Sielac-Olade.". Available online: https://sielac.olade.org/WebForms/Reportes/SistemaNumerico.aspx?ss=2 (accessed on 12 October 2023).
29. Secretaría de Energía. Resolución 240/2003. Available online: https://www.argentina.gob.ar/normativa/nacional/resoluci%25C3%25B3n-240-2003-87732/texto (accessed on 12 October 2023).
30. Ley de Electricidad Boliviana. 1994. Available online: https://www.elfec.bo/LEY1604#:~:text=La%2520ley%2520de%2520electricidad%2520Boliviana,en%2520todo%2520el%2520territorio%2520nacional (accessed on 12 October 2023).
31. Presidência da República. Law 9,648 of 1998. Available online: http://www.planalto.gov.br/ccivil_03/leis/L9648cons.htm (accessed on 12 October 2023).
32. DFL N° 1. 1982. Available online: https://www.bcn.cl/leychile/navegar?idNorma=3410 (accessed on 12 October 2023).
33. Ley 143 de 1994. Available online: https://www.funcionpublica.gov.co/eva/gestornormativo/norma.php?i=4631 (accessed on 12 October 2023).
34. General Electricity Law 2001. Available online: https://climate-laws.org/document/law-125-01-general-electricity-law_1683 (accessed on 12 October 2023).
35. Ley de Régimen del Sector Eléctrico. Available online: https://www.celec.gob.ec/transelectric/images/stories/baners_home/ley/Ley_de_regimen_del_sector_electrico.pdf (accessed on 12 October 2023).
36. SIGET. Available online: https://www.siget.gob.sv/las-telecomunicaciones-en-el-salvador/ (accessed on 12 October 2023).
37. CNEE. Available online: https://www.cnee.gob.gt/pdf/marco-legal/LEY%2520GENERAL%2520DE%2520ELECTRICIDAD%2520Y%2520REGLAMENTOS.pdf (accessed on 12 October 2023).
38. CREE. Available online: https://www.cree.gob.hn/wp-content/uploads/2019/02/LGIE-versi%C3%B3n-consolidada_CREE_jul_2022.pdf (accessed on 12 October 2023).
39. CENACE. Available online: https://www.cenace.gob.mx/Paginas/SIM/Liquidaciones.aspx (accessed on 12 October 2023).
40. Ley de la Industria Eléctricia. Ley No. 272. Available online: https://www.ine.gob.ni/?p=114585#:~:text=Establece%2520el%2520r%25C3%25A9gimen%2520legal%2520sobre,exportaci%25C3%25B3n%2520de%2520la%2520energ%25C3%25ADa%2520el%25C3%25A9ctrica (accessed on 12 October 2023).

41. Marco Regulatorio e Institucional Para la Prestación del Servicio Público de Electricidad. Available online: https://www.ecolex.org/es/details/legislation/ley-no-6-marco-regulatorio-e-institucional-para-la-prestacion-del-servicio-publico-de-electricidad-lex-faoc070278/#:~:text=derecho%2520ambiental-,Ley%2520N%25C2%25BA%25206%2520%252D%2520Marco%2520regulatorio%2520e%2520institucional%2520para%2520la,del%2520servicio%2520p%25C3%25BAblico%2520de%2520electricidad.&text=La%2520presente%2520Ley%2520establece%2520el,del%2520servicio%2520p%25C3%25BAblico%2520de%2520electricidad (accessed on 12 October 2023).
42. Decreto Ley No 25844. Available online: https://www.fao.org/faolex/results/details/es/c/LEX-FAOC064178/#:~:text=Per%25C3%25BA%2520(Nivel%2520nacional)-,Decreto%2520Ley%2520N%25C2%25BA%252025844%2520%252D%2520Ley%2520de%2520concesiones%2520el%25C3%25A9ctricas.,comercializaci%25C3%25B3n%2520de%2520la%2520energ%25C3%25ADa%2520el%25C3%25A9ctrica (accessed on 12 October 2023).
43. Ley N° 16832. Available online: https://www.impo.com.uy/bases/leyes/16832-1997 (accessed on 12 October 2023).
44. Charnes, A.; Cooper, W.W.; Rhodes, E. Measuring the efficiency of decision making units. *Eur. J. Oper. Res.* **1978**, *2*, 429–444. [CrossRef]
45. Mariz, F.B.; De Almeida, M.R.; Aloise, D. A review of Dynamic Data Envelopment Analysis: State of the art and applications. *Int. Trans. Oper. Res.* **2017**, *25*, 469–505. [CrossRef]
46. Tone, K.; Tsutsui, M. Dynamic DEA: A slacks-based measure approach. *Omega* **2010**, *38*, 145–156. [CrossRef]
47. Zhou, Z.; Liu, W. DEA models with undesirable inputs, intermediates, and outputs. In *International Series in Operations Research and Management Science*; Springer New York LLC.: New York, NY, USA, 2015; Volume 221, pp. 415–446. [CrossRef]
48. Shannon, C.E. A Mathematical Theory of Communication. *Bell Syst. Tech. J.* **1948**, *27*, 379–423. [CrossRef]
49. Meucci, A. Managing Diversification. *Risk* **2009**, *22*, 74–79.
50. Ruiz-Mendoza, B.J.; Sheinbaum-Pardo, C. Electricity sector reforms in four Latin-American countries and their impact on carbon dioxide emissions and renewable energy. *Energy Policy* **2010**, *38*, 6755–6766. [CrossRef]
51. Cragg, J.G. Some Statistical Models for Limited Dependent Variables with Application to the Demand for Durable Goods. *Econometrica* **1971**, *39*, 829–844. [CrossRef]
52. Kline, R.B. *Principles and Practice of Structural Equation Modelling*, 3rd ed.; Guilford Publications: New York, NY, USA, 2011.
53. Taylor, R. Interpretation of the Correlation Coefficient: A Basic Review. *J. Diagn. Med. Sonogr.* **1990**, *6*, 35–39. [CrossRef]
54. Olcay, M.; Laing, R. *Pharmaceutical Tariffs: What Is Their Effect on Prices, Protection of Local Industry and Revenue Generation?* WHO: Geneva, Switzerland, 2005.
55. Baldwin, R. Unilateral Tariff Liberalisation. *Int. Econ.* **2010**, *2010*, 10–43. [CrossRef]

*energies*

MDPI

*Article*

# Piezoelectric Sensors Pressed by Human Footsteps for Energy Harvesting

**Kyrillos K. Selim [1,*], Idris H. Smaili [2,*], Hossam M. Yehia [3], M. M. R. Ahmed [4] and Demyana A. Saleeb [5]**

1 Department of Electronics Technology, Faculty of Technology and Education, Helwan University, Cairo 11795, Egypt
2 Department of Electrical and Electronic Engineering, College of Engineering and Computer Science, Jazan University, P.O. Box114, Jazan 45142, Saudi Arabia
3 Department of Production Technology, Faculty of Technology and Education, Helwan University, Cairo 11795, Egypt; hossamelkeber@techedu.helwan.edu.eg
4 Department of Electrical Technology, Faculty of Technology and Education, Helwan University, Cairo 11795, Egypt; mohamedra1967@gmail.com
5 Faculty of Engineering, Kafrelsheikh University, Kafrelsheikh 33516, Egypt; demyanasaleeb@eng.kfs.edu.eg
* Correspondence: kyrillos.k.selim@techedu.helwan.edu.eg or kyrillos.beshara@hit.edu.cn (K.K.S.); ihsmaili@jazanu.edu.sa (I.H.S.)

**Abstract:** Human footsteps are a sustainable energy source that is derived from kinetic energy. As a result, in this study, piezoelectric sensors placed beneath floor tiles were excited by human footsteps to provide practical electrical energy. A simple rectifying circuit with a filter was used to capture electrical power. The floor tile is 455 mm in length and 405 mm in width. Two light-emitted diodes were lit up as the actual load by utilising electrical energy obtained from the kinetic energy generated by human footsteps. The greatest attainable power that could be extracted from the suggested floor tile was 249.6 milliwatts, with an approximate cost of $10.2.

**Keywords:** energy harvesting; DC-DC converter; floor tile; piezoelectric generator; Schottky diode

## 1. Introduction

There is an increasing demand for daily electrical energy due to the emergence of electrical and electronic devices that rely on electricity. There are clean renewable energy sources, such as sound energy, vibration energy, wind energy, sea wave energy, and kinetic energy. Proper exploitation of such sources helps reduce the burden of the increasing demand for traditional energy sources while at the same time benefiting from sources that are permanently and freely available. Human footsteps are available where people live. Furthermore, human footsteps are a waste of energy. Given the global scarcity of energy resources, it is imperative to actively seek alternative sources, such as harnessing the power generated by human footsteps, rather than considering it a choice. Piezoelectric transduction is an up-and-coming technology for harvesting ambient energy. It has been successfully applied in various fields such as structures, transportation, wireless electronics, microelectromechanical systems (MEMS), the Internet of Things (IoT), and wearable and implantable biomedical devices. Piezoelectric materials are used in one of these energy harvesting methods to transform kinetic energy or mechanical vibrations in the surrounding environment into electrical power. Pedestrian energy harvesting utilises the piezoelectricity phenomenon to convert the kinetic energy created by individuals walking, jogging, and leaping into electricity. This technology has the potential to power a wide range of applications, from powering small sensors to illuminating streets, to even charging electric vehicles as a future dream [1,2].

Researchers have used a variety of designs to embed piezoelectric components for mechanical-to-electric energy conversion in floor tiles. These designs are roughly categorised into three primary types: cantilever type [1–8], curved type [9–11], and array/stacked type [12–22], depending on the type of piezoelectric element utilised. However, the most readily accessible piezoelectric energy harvesters are piezoelectric diaphragms, which are simple to attach to any planar surface using an adhesive. This kind of element in piezoelectric tiles has been the subject of several research studies [23].

Regarding the electrical design, different techniques have been adopted. Chew et al. [24] showed an adaptive self-configuring rectifier that uses MOSFET tubes to switch between the voltage doubler and the full-bridge rectifier. This technique allows for the harvesting of lower-voltage charges. Lefeuvre et al. [25] suggested a synchronous electric charge extraction interface based on the chopper circuit, which uses a nonlinear method to increase the harvested electrical power. Hong Jiang et al. [26] proposed a circuit to control the extraction of electric charges. The circuit is composed of an energy-harvesting circuit and a control circuit. The control circuit was used to create control signals for the switch to turn on and off, allowing the energy-harvesting circuit portion to function correctly.

Based on the aforementioned details, some studies have focused on the mechanical design of pedestrian energy harvesters [5,6,27,28] to efficiently convert pedestrian pressure into electrical power. Other studies have concentrated on electrical design to ensure stable management of the harvested power [26,29]. However, some literature has addressed mechanical and electrical designs [7,30,31].

It is possible to generate higher power by connecting multiple piezoelectric energy harvesters. However, piezoelectric energy harvester cannot be directly connected to electrical loads because the high output voltage changes based on how people walk and how much pressure there is over time. Furthermore, a simple full-wave bridge rectifier is insufficient because of the unstable waveform that results from the rectification stage. As a result, a suitable control circuit is needed to solve these problems [7,26]. This study examines the practicality of producing power from piezoelectric crystals and provides a detailed plan for the mechanical and electrical components of the proposed energy harvesting system. Furthermore, it incorporates experimental validation to assess the mechanical aspects. The manuscript is organised as follows: Section 2 is the Design and Experimental Validation, Section 3 presents the Results and Discussion, and Section 4 covers the Conclusion of this study.

## 2. Design and Experimental Validation

The structure has three plates placed in a sandwich-like form. Plate No. 1 depicts the frontal plate, measuring 455 mm × 405 mm × 15 mm, as illustrated in Figure 1. Plate No. 2 is a substrate plate with dimensions of 455 mm × 405 mm × 3 mm. Plate No. 3 is positioned between plates 1 and 2, with dimensions of 420 mm × 380 mm × 12 mm.

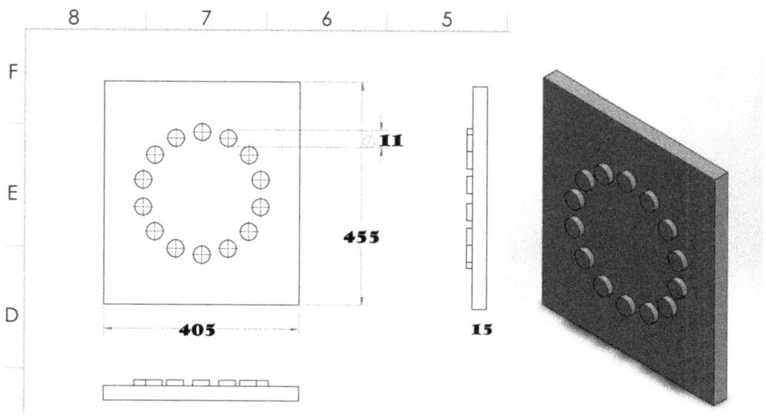


(a) (b)

**Figure 1.** Dimensions of wood plate No. 1 (**a**): different views and (**b**) 3D wood plate.

Fourteen hot melt glue sticks, each measuring 10 mm in length, were evenly spaced on plate No. 1 to form a circle with a circumference of 300 mm, as shown in Figure 1. The hot-melt adhesive sticks were used to apply direct pressure to each piezoelectric sensor in the cavity whenever there was movement from human footsteps on the floor tile. The technique employed hot-melt adhesive sticks and foam materials that possessed sufficient flexibility to endure elevated pressures.

Our experiments used a ceramic piezoelectric diaphragm with a circular form. This sensor contains certain technical features, as follows: The metal plate size diameter is 20 mm, the piezoelectric element size diameter is 14 mm, the electrode size diameter is 12.8 mm, the piezoelectric material thickness is 0.42 mm, the metal plate thickness is 0.2 mm, and the metal plate material is composed of brass with lead wire. Figure 2 depicts the circuit that is equivalent to a piezoelectric element. The piezoelectric current ($I_P$) and voltage ($V_P$) can be expressed using Equations (1) and (2), respectively. In these equations, $R_P$ represents the piezoelectric resistance, $C_P$ represents the piezoelectric capacitance, and $\omega$ represents the angular frequency.

$$V_P = I \cdot \frac{\frac{1}{J\omega C_P} \cdot R_P}{\frac{1}{J\omega C_P} + R_P} = I \cdot \frac{R_P}{1 + J\omega C_P R_P} \quad (1)$$

$$I_P = I - V_P(J\omega C_P) - \frac{V_P}{R_P} \quad (2)$$

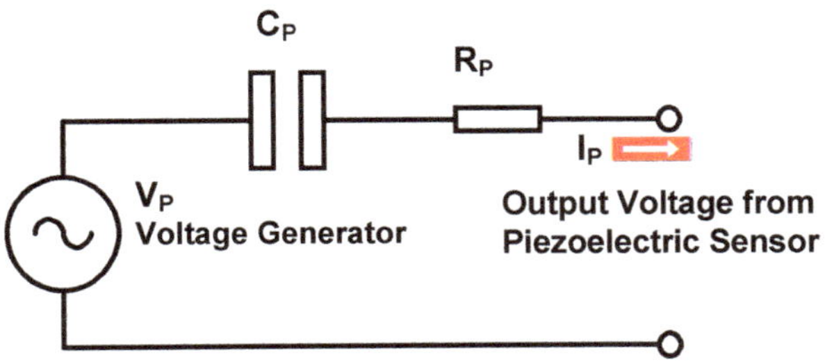


**Figure 2.** Piezoelectric equivalent circuit.

The 14 piezoelectric generators connected in parallel can be represented by the equivalent current ($I_{P14}$) and voltage ($V_{P14}$), which are provided in Equations (3) and (4) respectively, as shown in Figure 2.

$$V_{P14} = I \cdot \frac{\frac{1}{J\omega C_P} \cdot \frac{R_P}{14}}{\frac{1}{J\omega C_P} + \frac{R_P}{14}} = I \cdot \frac{R_P}{1 + J\omega C_P R_P} = V_P \quad (3)$$

$$I_{P14} = 14I - V_P(14J\omega C_P) - \frac{14V_P}{R_P} \quad (4)$$

Figure 1 illustrates the distribution of the fourteen piezoelectric harvesters around the rectangular floor tile, whereas Figure 3 shows the experimental configuration of the suggested footstep energy harvesting system. In this study, the researchers utilised the bridge rectifier circuit and opted for the 1N60P Schottky diode. The choice was made based on the diode's low forward voltage of 0.24 V under low current conditions, which is significantly lower compared to silicon diodes. Table 1 presents the key properties of the 1N60P Schottky diode.

The rectifier circuit consisted of four Schottky diodes of the 1N60P type, which are linked in conjunction with a 2.2 µF filter capacitor, as illustrated in Figure 4.

**Table 1.** 1N60P diode Characterestics.

| Symbol | Parameter | Condition | Value | | Unit |
|---|---|---|---|---|---|
| | | | Typ. | Max. | |
| VRRM | Repetitive Peak Reverse Voltage | – | 45 | – | V |
| $I_F$ | Forward Continuous Current | $T_A$ = 25 °C | 50 | – | mA |
| $V_F$ | Forward Voltage | $I_F$ = 1 mA | 0.24 | 0.5 | V |
| $I_R$ | Reverse Current | $V_R$ = 15 V | 5 | 10 | μA |
| $C_J$ | Junction Capacitance | $V_R$ = 10 V & F = 1 MHz | 6 | – | pF |

(a)

(b) (c)

**Figure 3.** The experimental configuration of the suggested footstep energy harvesting system consists of three main components: (**a**) wood plate No. 1; (**b**) piezoelectric elements mounted on foam and wood substrate plates; and (**c**) plates No. 1, 2, and 3.

The rectified output voltage, denoted as ($V_{out}$), is determined by Equation (5). The average voltage ($V_{avg}$) can be determined using Equations (6) and (7).

$$V_{out} = V_P sin(\omega t); 0 \leq \omega \leq \pi \tag{5}$$

$$V_{avg} = \frac{1}{\pi} \int_0^{2\pi} V_{out} d(\omega t) \tag{6}$$

$$V_{avg} = \frac{1}{\pi} \int_0^{2\pi} V_P sin(\omega t) d(\omega t) \tag{7}$$

where $t$ is the variable for time. The formulas for the instantaneous power ($P_t$) and average power ($P_{avg}$) in the bridge rectifier are provided in Equations (8) and (9), respectively.

$$P_{out} = \frac{V(t)^2}{R_L} \tag{8}$$

$$P_{avg} = \frac{1}{R_L} \cdot \frac{1}{T} \int_0^T P(t)dt \tag{9}$$

The insertion of a filter capacitor ($C_F$) alongside the bridge rectifier results in the expected DC output voltage ($V_{dc}$) across the load ($R_L$) and the harvested power ($P_{dc}$), as indicated by Equations (10) and (11).

$$V_{dc} = \left[1 - \frac{1}{2fR_LC_F}\right]\left[\frac{V_m}{2} - 0.7\right] \tag{10}$$

$$P_{dc} = \frac{Vdc^2}{R_L} \tag{11}$$

As $f$ is the frequency.

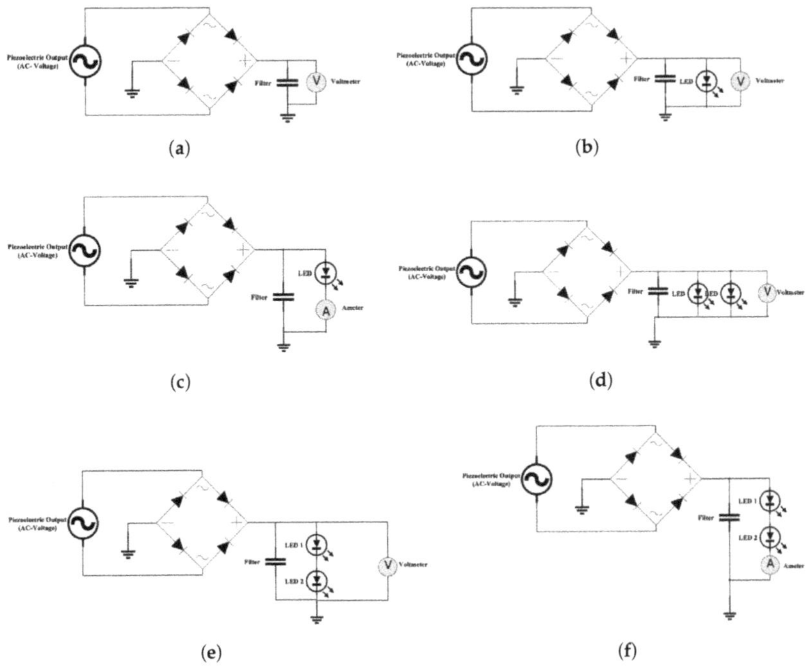


**Figure 4.** Quantitative assessments of various states of bridge rectifier circuits: The measurements include: (**a**) open loop voltage measurement, (**b**) single load voltage measurement, (**c**) single load current measurement, (**d**) two parallel load voltage measurement, (**e**) two series load voltage measurement, and (**f**) two series load current measurement.

The output voltage level produced by the typical rectifier and filter was not stable due to the random pressure of humans on the energy-harvesting floor tile. Moreover, this rectified voltage varied from 0 V to around 20 V in the open circuit condition. A DC–DC buck converter with a control system was needed to export a stable voltage to the electrical load (the desired application). The DC–DC buck converter was adopted in this work to step down the high voltage generated by piezoelectric generators to fit low-power applications such as wireless sensors. In addition, the two-loop cascade control was utilized to fix the output voltage at a constant level (5 V). The proposed control system was built on the Simulink platform. Moreover, as the piezoelectric sensor is not listed in the library of Simulink, the electrical equivalent circuitof the piezoelectric sensor was built and utilised in simulations. The piezoelectric sensor was represented by a voltage source connected in series with a resistor ($R_P$) and internal capacitance ($C_P$), as shown in Figures 2 and 5. The values of $C_P$ and $R_P$ were extracted from the piezoelectric sensor datasheet. However, due to the stochastic nature of the energy-harvesting process of human footsteps, this frequency depends on the situation and the place where the system is applied. For the Simulink model, we used 10 Hz as an average value.

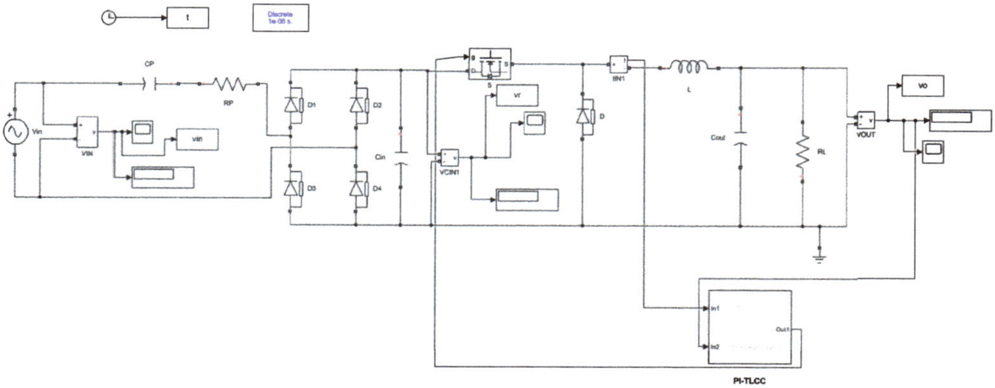


**Figure 5.** The piezoelectric energy harvesting system model in Simulink with the control circuit.

Two cases were considered regarding the input voltage variation of the voltage generator ($V_P$) in this study; the first case handled the voltage variation from −20 V to +20 V, whereas the second case handled the voltage variation from −10 V to +10 V.

The system proposed in this study involved a series equivalent circuit for piezoelectric, connected to a buck converter, and controlled using the Two-Loop Cascade Control (TLCC) method. This system is illustrated in Figure 5. The alternating voltage was rectified into a direct current (DC) voltage by the use of a full-wave rectifier. The value was subsequently rendered uniform by employing a filter capacitor, leading to a stable direct current (DC) value. The DC-DC converter step, specifically a buck converter, was tasked with reducing the input voltage to 5 V. The stable and uniform output was obtained by adopting a two-loop cascade control (TLCC) system.

The suggested control system for the DC-DC buck converter consists of two loops: the voltage outer loop and the current inner loop. The system produces an inductor reference current by comparing the output voltage with a voltage reference. The inner loop calculates the duty cycle for the pulse width modulation. Figure 5 illustrates the suggested PI-TLCC Simulink model for the DC–DC buck converter. The control system's objective is to accurately follow the reference signal, which is an intended external signal provided to the outer control loop. The outer loop controller is a proportional-integral (PI) controller that adjusts the output voltage of the converter. The inner loop consists of a proportional-integral (PI) controller that is specifically intended to regulate the current flowing through the inductor [32,33]. The controller gains were computed using Ziegler–Nichol's approach to ensuring precise tuning of the PI controller parameters [34]. The formulae for the reference current and controller output ($U$) in the PI-TLCC approach are provided in Equations (12) and (13) correspondingly.

$$I_{ref} = (V_{ref} - V_{out})(K_p + \frac{K_i}{S}) \tag{12}$$

$$U = (I_{ref} - I_l)(K_p + \frac{K_i}{S}) \tag{13}$$

where $K_p$ and $K_i$ are the PI control gain parameters. $V_{ref}$ reference voltage $V_{out}$ output voltage $I_{ref}$ reference current $I_l$ inductor current

The purpose of using this method is to stabilise the output value at 5 V when any changes or fluctuations occur in the input voltage or the applied load.

The system output depends on the applied load, and in this article, there is a linear load consisting of resistors whose value changes from 100 ohms to 200 ohms. In order to enhance the converter's performance, the Ziegler–Nichols approach was applied in this design to ascertain the control parameters, namely the proportional and integral gains. The buck converter was tuned to have a switching frequency of 40 kHz. In addition, a MOSFET was employed as a switch due to its capability of high-speed switching at very high frequencies. It is widely used in switch-mode power supply (SMPS) circuits and provides a unique advantage.

*Parameters of the Buck Converter*

1. Input voltage range: $V_{In-Min}$ and $V_{In-Max}$
2. Rated output voltage: $V_{Our}$
3. Maximum output current: $I_{Our-Max}$

The duty cycle $(D)$ of the maximum input voltage is determined first before calculating the switch current. The maximum input voltage was selected because it produces the highest switch current.

$$D = \frac{V_{Out}}{V_{In-Max} \cdot \eta} \tag{14}$$

where $V_{In-Max}$ is the maximum input voltage, $V_{Out}$ is the desired output voltage, and $\eta$ is the converter efficiency. The maximum switch current is then calculated by determining the inductor ripple current.

$$\Delta I_L = \frac{(V_{In-Max} - V_{Out}) \cdot D}{f_S \cdot L} \tag{15}$$

where $f_s$ is the converter's minimum switching frequency, and $L$ is the specified inductor value. For components with no inductor range specified, the following equation is a fair estimate for the appropriate inductor:

$$L = \frac{V_{Out} \cdot (V_{In} - V_{Out})}{\Delta I_L \cdot f_S \cdot V_{In}} \tag{16}$$

where $V_{IN}$ is the typical input voltage and $\Delta_{IL}$ is the estimated inductor ripple current. An accurate estimate for the inductor ripple current is 20% to 40% of the output current.

$$\Delta I_L = 0.2to0.4 \cdot I_{Out-Max} \tag{17}$$

$I_{Out-Max}$ is the maximum output current required in the application. The following formula could be applied to change the output capacitor values to obtain the appropriate output voltage ripple with external compensation:

$$C_{Out-Min} = \frac{\Delta I_L}{8 \cdot f_S \cdot \Delta V_{Out}} \tag{18}$$

$C_{Out-Min}$ is the minimum output capacitance, and $\Delta V_{OuT}$ is the required output voltage ripple. The output capacitor's ESR adds slightly more ripple, as shown by the equation below:

$$\Delta V_{Out}(ESR) = ESR \cdot \Delta I_L \tag{19}$$

$\Delta V_{OuT(ESR)}$ is the additional output voltage ripple caused by capacitors' ESR, and ESR is the output capacitor's equivalent series resistance.

The energy harvesting scheme simulated in Simulink is pictured in Figure 5. The system consisted of four parts. The first part is the piezoelectric equivalent circuit, which consists of the AC voltage source (A = 20 v, F = 10 Hz), a first capacitor (10 nF), and a resistor (1000 Ω) in series. The second part is the rectifier bridge, which consists of four diodes and a parallel capacitor (3000 µF). The third part is the buck converter, with a reference voltage of 5 V. The last part is the load, represented as a parallel resistor (100 Ω).

## 3. Results and Discussion

The output voltage and current of the bridge rectifier circuit were measured and recorded under different conditions. Table 2 presents the most noteworthy findings. In relation to the open-loop state, the bridge rectifier circuit produced a significantly elevated output voltage, as depicted in Table 2. The bridge rectifier experienced a peak voltage of 33.7 V.

Moreover, when the LED was employed as a practical load, whether with one LED (a single load) or two LEDs (two loads connected in series or in parallel), the voltage generated at the terminals of the load was equal to the LED's operating voltage. This suggests that the generated voltage was adequate to supply power to both single and two loads.

**Table 2.** Quantitative data obtained from applying pressure from a single footstep on the tile using the rectifier.

| Circuit Test Condition | Quantity | Measured | Unit |
|---|---|---|---|
| Open loop condition | Output Voltage | 33.7 | V |
| 1 LED as a load | Output Voltage | 2.4 | V |
| 1 LED as a load | Output Current | 0.079 | *A* |
| 2 series LEDs as a load | Output Voltage | 4.8 | V |
| 2 parallel LEDs as a load | Output Voltage | 2.4 | V |

Figure 6 illustrates the practical load measurement (at the condition of a single load). The output voltage across the load of two series LEDs varied between 4.6 V and 4.8 V at a speed of around 80 pedestrians per minute. The generated current was recorded as 0.052 A, whereas the maximum harvested power was 249.6 mW.

Moreover, an AC input voltage of 10 volts was used as the maximum value, resulting in a rectifier voltage of 8.4 V and an output voltage of 5 V, as shown in Figures 7–9. When a maximum input voltage of 20 V was applied, the voltage outside the bridge was 18.4 V, and the output voltage remained at 5 V, as shown in Figures 10–12. We used a linear load, such as resistors with a 100 to 200 ohms value, and the results were satisfactory. Overall, a two-loop cascade control was adopted in this study, which improved the output voltage level and resulted in a stable voltage waveform.

Simulation parameter specifications of the buck converter are listed in Table 3. The simulation of the proposed system was done in 2 s.

**Table 3.** Simulation parameter specifications of a buck converter.

| Parameter | Value |
|---|---|
| $VAC_{max}$ | 10 V and 20 V |
| $L$ | 740 µH |
| $C$ | 100 µF |
| $R_L$ | 100:200 Ω |
| $V_{Out}$ | 5 V |
| Switching Frequency | 40 kHz |
| Piezoelectric excitation frequency | 10 Hz |

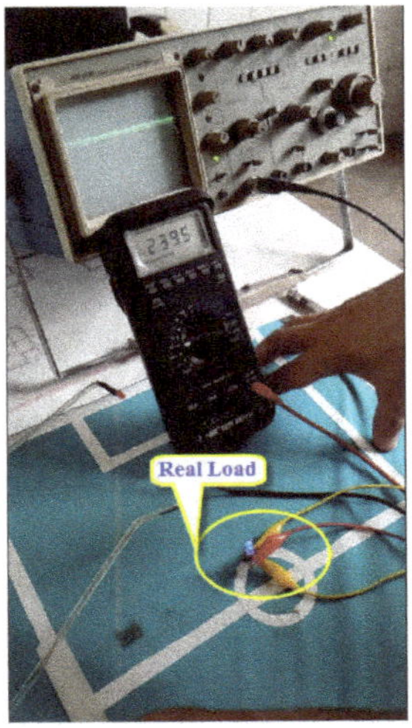


**Figure 6.** The voltage doubler circuit is coupled to the multimeter and oscilloscope to measure the practical load.

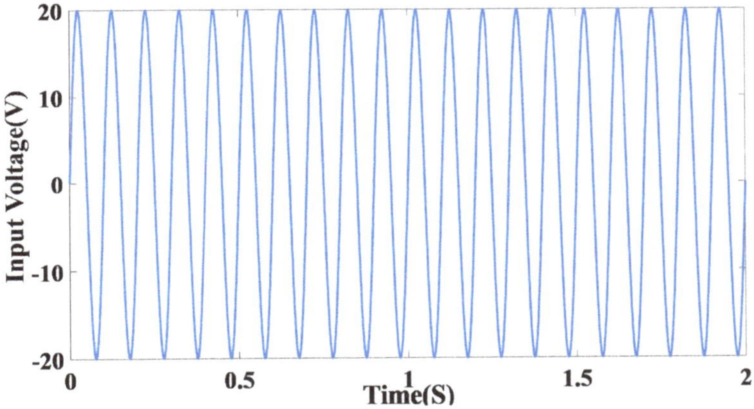


**Figure 7.** Input voltage waveform produced by the piezoelectric array for case No. 1.

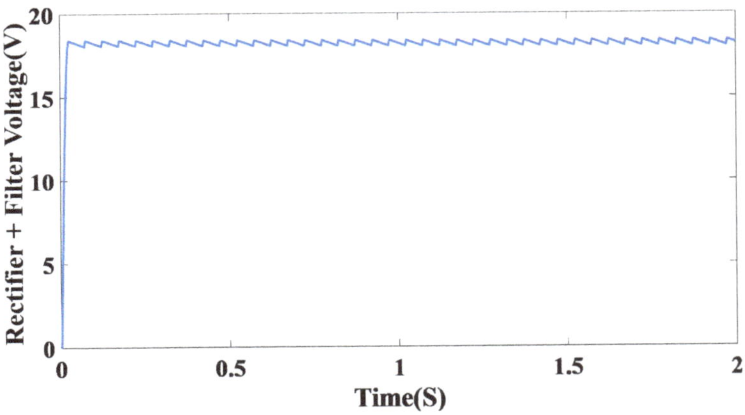


**Figure 8.** Output voltage waveform produced by the bridge and filter for case No. 1.

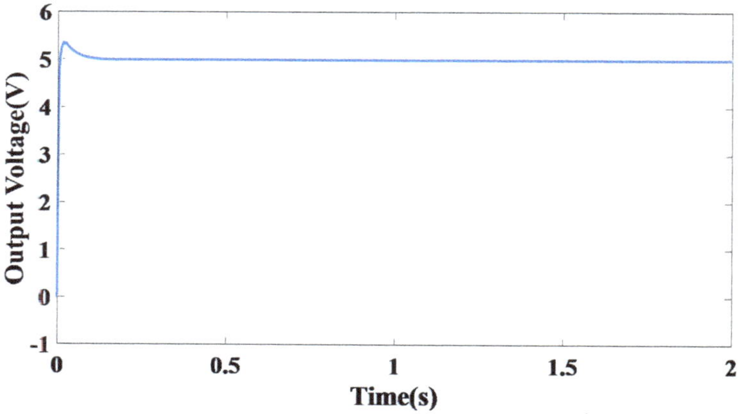


**Figure 9.** Output voltage waveform produced by the control system for case No. 1.

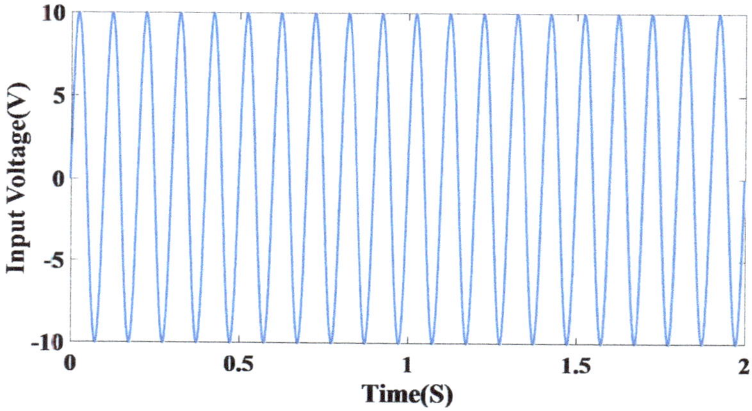


**Figure 10.** Input voltage waveform produced by the piezoelectric array for case No. 2.

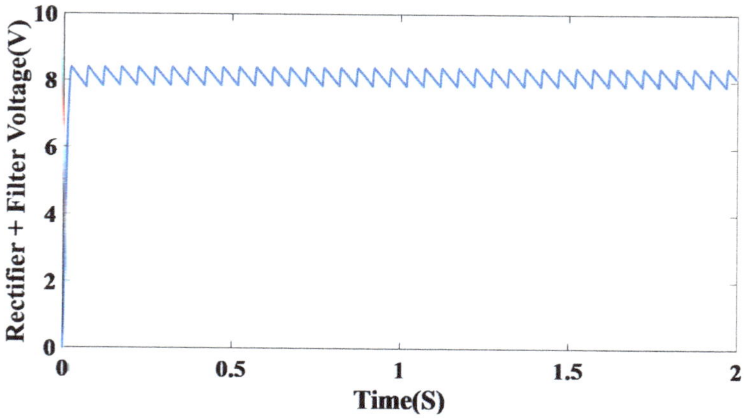


**Figure 11.** Output voltage waveform produced by the bridge and filter for case No. 2.

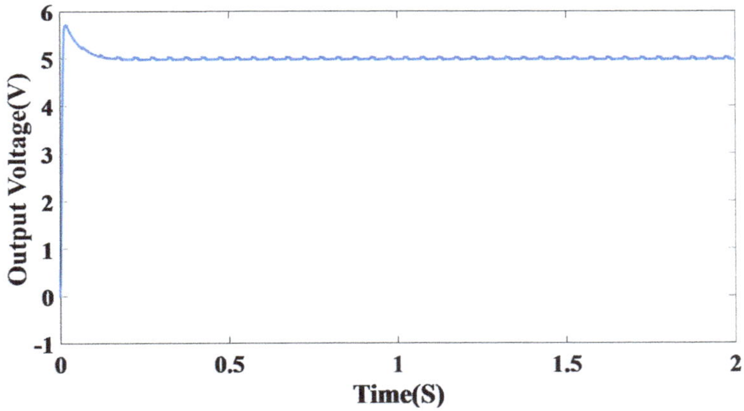


**Figure 12.** Output voltage waveform produced by the control system for case No. 2.

The expected energy from one tile during one hour is

$$E(J) = P(W) \times t(s) = 250 \text{ mW} \times 60 \times 60 = 900 \text{ J} = 0.25 \text{ WHr}. \tag{20}$$

The expected energy from one tile during 8 h is

$$E(J) = 8 \times 900 = 7200 \text{ J} \ = 2 \text{ WHr} \tag{21}$$

The entrance gate to the college building is 8 $m^2$, so 30 tiles can be installed in the entrance gate. Therefore, the expected energy generated by 30 tiles within an hour is

$$E = 0.25 \times 30 = 7.5 \text{ WHr} \tag{22}$$

Hence, the expected energy from 30 tiles in 8 h is

$$E = 7.5 \times 8 = 60 \text{ WHr} \tag{23}$$

A security CCTV camera, model TP-Link Tapo C310, has a power consumption of 5.4 watts and can be installed on the entrance gate of the college building for security purposes. So, the energy harvested can operate the security CCTV camera for around 11 h. The average number of students per day in the faculty is around 2000. The energy harvesting technology, once adopted, has a price of approximately $10.2 per individual tile.

Finally, a comparison with similar methods of using piezoelectric elements to generate energy and this study were illustrated in Table 4 concerning the harvester type, floor tile dimensions, and amount of harvested power.

**Table 4.** An analysis of the proposed work in relation to existing literature

| Reference | Harvester Type | Tile Size (mm) | Power Produced (mW) |
|---|---|---|---|
| [14] | Piezoelectric | 300 × 4300 | 6.04 |
| [3] | PZN 0.5C Thick Film Cantilever | 200 × 200 | 12 |
| [4] | PZT Cantilever | 430 × 430 | 1.24 |
| [28] | PZT Cantilever | 450 × 450 | 35 |
| Proposed work | Piezoelectric Ceramic Diaphragm | 455 × 405 | 249.6 |

## 4. Conclusions

A piezoelectric generator-based energy harvesting device was developed that relies on human footsteps. The bridge rectifier achieved a peak output voltage of 33.7 V. The prototype could light up two LEDs and had dimensions of 455 mm in length and 405 mm in width. Further, the proposed control achieved the stability of the output voltage of the harvested power at a level of 5 V over any variation at the input due to the stochastic nature of the mechanical excitation source. An assembly of 30-floor tiles can be placed at the entrance gate of the Faculty of Technology at Helwan University, which is anticipated to generate 60 watt-hours of energy over 8 h.

**Author Contributions:** Conceptualization, K.K.S., I.H.S., H.M.Y., M.M.R.A. and D.A.S.; methodology, K.K.S.; software, K.K.S. and H.M.Y.; validation, K.K.S., I.H.S., H.M.Y., M.M.R.A. and D.A.S.; formal analysis, K.K.S., I.H.S., H.M.Y., M.M.R.A. and D.A.S.; investigation, K.K.S., I.H.S., H.M.Y., M.M.R.A. and D.A.S.; resources, K.K.S., I.H.S., H.M.Y., M.M.R.A. and D.A.S.; data curation, K.K.S., I.H.S., H.M.Y., M.M.R.A. and D.A.S.; writing—original draft preparation, K.K.S., I.H.S., H.M.Y., M.M.R.A. and D.A.S.; writing—review and editing, K.K.S., I.H.S., H.M.Y., M.M.R.A. and D.A.S.; visualization, K.K.S., I.H.S., H.M.Y., M.M.R.A. and D.A.S. All authors have read and agreed to the published version of the manuscript.

**Funding:** This research received no external funding.

**Data Availability Statement:** The original contributions presented in the study are included in the article, further inquiries can be directed to the corresponding author.

**Conflicts of Interest:** The authors declare no conflicts of interest.

## References


1. Hwang, S.J.; Jung, H.; Kim, J.; Ahn, J.; Song, D.; Song, Y.; Lee, H.; Moon, S.; Park, H.; Sung, T. Designing and manufacturing a piezoelectric tile for harvesting energy from footsteps. *Curr. Appl. Phys.* **2015**, *15*, 669–674. [CrossRef]
2. El-Etriby, A.E.; Abdel-Meguid, M.E.; Hatem, T.M.; Bahei-El-Din, Y.A. A multiscale-based approach for composite materials with embedded PZT filaments for energy harvesting. In *Behavior and Mechanics of Multifunctional Materials and Composites*; SPIE: Bellingham, WA, USA, 2014; Volume 9058, pp. 294–305.
3. Kim, K.-B.; Cho, J.Y.; Jabbar, H.; Ahn, J.; Hong, S.; Woo, S.B.; Sung, T. Optimized composite piezoelectric energy harvesting floor tile for smart home energy management. *Energy Convers. Manag.* **2018**, *171*, 31–37. [CrossRef]
4. Panthongsy, P.; Isarakorn, D.; Janphuang, P.; Hamamoto, K. Fabrication and evaluation of energy harvesting floor using piezoelectric frequency up-converting mechanism. *Sens. Actuators A Phys.* **2018**, *279*, 321–330. [CrossRef]
5. Panthongsy, P.; Isarakorn, D.; Hamamoto, K.; Janphuang, P. Performance and behavior analysis of piezoelectric energy harvesting floor tiles. In Proceedings of the 2019 5th International Conference on Engineering, Applied Sciences and Technology (ICEAST), Luang Prabang, Laos, 2–5 July 2019; pp. 1–4.
6. Isarakorn, D.; Jayasvasti, S.; Panthongsy, P.; Janphuang, P.; Hamamoto, K. Design and evaluation of double-stage energy harvesting floor tile. *Sustainability* **2019**, *11*, 5582. [CrossRef]
7. Jabbar, H.; Hong, S.D.; Hong, S.K.; Yang, C.H.; Jeong, S.Y.; Sung, T.H. Sustainable micro-power circuit for piezoelectric energy harvesting tile. *Integr. Ferroelectr.* **2017**, *183*, 193–209. [CrossRef]
8. Jhun, J.P.; Hong, S.D.; Jeon, D.H.; Sung, T.H. The effectiveness of different width piezoelectric energy harvester in the pedestrian floor tile energy harvesting system for internet of things sensors. *J. Korean Phys. Soc.* **2021**, *78*, 81–88. [CrossRef]
9. Liu, Z.; Ding, G.; Wang, J.; Cai, G.; Qin, X.; Shi, K. Fabrication and performance of Tile transducers for piezoelectric energy harvesting. *AIP Adv.* **2020**, *10*, 045326. [CrossRef]
10. Kathpalia, B.; Tan, D.; Stern, I.; Valdes, F.; Kim, S.; Erturk, A. Modeling and characterization of a curved piezoelectric energy harvester for smart paver tiles. *Procedia Comput. Sci.* **2017**, *109*, 1060–1066. [CrossRef]
11. Wang, J.; Cai, Y.; Liu, Z.; Ding, G.; Cai, G.; Fu, H. Preparation and performance study of a new type of Tile transducer for roadway applications. *J. Intell. Mater. Syst. Struct.* **2020**, *31*, 2020–2028. [CrossRef]
12. Kamal, P.N.M.; Buniyamin, N. Using piezoelectric elements as footsteps energy harvester: An investigation. In Proceedings of the 2018 IEEE 8th International Conference on System Engineering and Technology (ICSET), Bandung, Indonesia, 15–16 October 2018; pp. 1–6.
13. Puscasu, O.; Counsell, N.; Herfatmanesh, M.R.; Peace, R.; Patsavellas, J.; Day, R. Powering Lights with Piezoelectric Energy-Harvesting Floors. *Energy Technol.* **2018**, *6*, 906–916. [CrossRef]
14. Abadi, P.B.; Darlis, D.; Suraatmadja, M.S. Green energy harvesting from human footsteps. *MATEC Web Conf.* **2018**, *197*, 11015. [CrossRef]
15. Cascetta, F.; Schiavo, A.L.; Minardo, A.; Musto, M.; Rotondo, G.; Calcagni, A. Analysis of the energy extracted by a harvester based on a piezoelectric tile. *Curr. Appl. Phys.* **2018**, *18*, 905–911. [CrossRef]
16. Saifan, R.; Ali, L.A.; Shreikh, A.A.; Alnabelsi, S.H. Smart walk: Case studies on hybrid power generation system of piezoelectricity and solar power. In Proceedings of the 2019 International Conference on Electrical and Computing Technologies and Applications (ICECTA), Ras Al Khaimah, United Arab Emirates, 19–21 November 2019; pp. 1–6.

17. Covaci, C.; Porobic, I.; Gontean, A. Setup for Piezoelectric Energy Harvesting System. In Proceedings of the 2019 IEEE 25th International Symposium for Design and Technology in Electronic Packaging (SIITME), Cluj-Napoca, Romania, 23–26 October 2019; pp. 78–81.
18. Ruman, M.R.; Das, M.; Mahmud, S.I. Human footsteps for energy generation by using piezoelectric tiles. In Proceedings of the 2019 Innovations in Power and Advanced Computing Technologies (i-PACT), Vellore, India, 22–23 March 2019; Volume 1, pp. 1–6.
19. Ahmed, S.U.; Sabir, A.; Ashraf, T.; Ali Haider, M.; Perveen, F.; Farooqui, Z.; Uddin, R. Energy Harvesting through Floor Tiles. In Proceedings of the 2019 International Conference on Innovative Computing (ICIC), Lahore, Pakistan, 1–2 November 2019; pp. 1–6.
20. Selim, K.K.; Haggag, A.; Amer, F.A.; Rady, W.; El-Garhy, A.M. A proposed technique for power extraction from acoustic energy scavenging. *Int. J. Electron.* **2018**, *105*, 1236–1247. [CrossRef]
21. Selim, K.K.; Haggag, A.; Yehia, H.; Amer, F.Z. Acoustic energy conversion into useful electric energy from Disk Jockey by using piezoelectric transducers. In Proceedings of the 2016 Eighteenth International Middle East Power Systems Conference (MEPCON), Cairo, Egypt, 27–29 December 2016; pp. 871–876.
22. Selim, K.K.; Yehia, H.M.; Abdalfatah, S. Human footsteps-based energy harvesting using piezoelectric elements. In Proceedings of the 2023 International Telecommunications Conference (ITC-Egypt), Alexandria, Egypt, 18–20 July 2023.
23. Sharma, S.; Kiran, R.; Azad, P.; Vaish, R. A review of piezoelectric energy harvesting tiles: Available designs and future perspective. *Energy Convers. Manag.* **2022**, *254*, 115272. [CrossRef]
24. Chew, Z.J.; Zhu, M. Adaptive self-configurable rectifier for extended operating range of piezoelectric energy harvesting. *IEEE Trans. Ind. Electron.* **2019**, *67*, 3267–3276. [CrossRef]
25. Lefeuvre, E.; Badel, A.; Richard, C.; Guyomar, D. Piezoelectric energy harvesting device optimization by synchronous electric charge extraction. *J. Intell. Mater. Syst. Struct.* **2005**, *16*, 865–876. [CrossRef]
26. Jiang, H.; Wang, Y.; Wu, G.; Li, B.; Li, X. A novel control method of electric charge extraction circuit for piezoelectric energy harvesters. *Ferroelectrics* **2022**, *599*, 58–71. [CrossRef]
27. Fu, H.; Xu, R.; Seto, K.; Yeatman, E.; Kim, S. Energy harvesting from human motion using footstep-induced airflow. *J. Phys. Conf. Ser.* **2015**, *660*, 012060. [CrossRef]
28. Yingyong, P.; Thainiramit, P.; Jayasvasti, S.; Thanach-Issarasak, N.; Isarakorn, D. Evaluation of harvesting energy from pedestrians using piezoelectric floor tile energy harvester. *Sens. Actuators A Phys.* **2021**, *331*, 113035. [CrossRef]
29. Abidin, N.A.K.Z.; Nayan, N.M.; Ali, A.; Azli, N.; Nordin, N. Analysis of AC-DC Converter Circuit Performance with Difference Piezoelectric Transducer Array Connection. *Int. J. Integr. Eng.* **2020**, *12*, 20–27.
30. Chand, A.A.; Arefin, A.S.M.S.; Islam, F.; Prasad, K.; Singh, S.; Cirrincione, M.; Mamun, K. Design simulation of a novel fluid based footstep energy harvesting system. *Sustain. Energy Technol. Assess.* **2020**, *39*, 100708. [CrossRef]
31. Jeong, S.Y.; Cho, J.Y.; Hong, S.D.; Hwang, W.; Jabbar, H.; Ahn, J.H.; Jhun, J.P.; Sung, T.H. Self-powered operational amplifying system with a bipolar voltage generator using a piezoelectric energy harvester. *Electronics* **2019**, *9*, 41. [CrossRef]
32. Galeano, N.M.; Lezama, J.M.L.; Quintero, J.B.C. A Methodology for Tunning Cascade PI Controllers for Power Electronics Converters. *Int. J. Eng. Res. Technol.* **2020**, *13*, 3997–4003. [CrossRef]
33. Azarastemal, S.; Hejri, M. Cascade Control System Design and Stability Analysis for a DC–DC Boost Converter with Proportional Integral and Sliding Mode Controllers and Using Singular Perturbation Theory. *Iran. J. Sci. Technol. Trans. Electr. Eng.* **2021**, *45*, 1445–1462. [CrossRef]
34. Mohanty, P.R.; Panda, A.K. Fixed-frequency sliding-mode control scheme based on current control manifold for improved dynamic performance of boost PFC converter. *IEEE J. Emerg. Sel. Top. Power Electron.* **2016**, *5*, 576–586. [CrossRef]

MDPI AG
Grosspeteranlage 5
4052 Basel
Switzerland
Tel.: +41 61 683 77 34

*Energies* Editorial Office
E-mail: energies@mdpi.com
www.mdpi.com/journal/energies

www.ingramcontent.com/pod-product-compliance
Lightning Source LLC
Chambersburg PA
CBHW061556210926
51850CB00009B/2124
* 9 7 8 3 7 2 5 8 6 8 4 2 1 *